AGRICULTURAL RESEARCH AND DEVELOPMENT

ABOUT THE AUTHOR

Dr. Satya Ranjan Barooah is an outstanding Agricultural Scientists-cum-Administrator. An eminent Technocrat.

He passed his B.Sc. (Ag.) from Allahabad Agricultural Institute under Allahabad University, M.Sc. (Ag.) from the Hindu University in first class standing 2nd Ph.D. from Allahabad University and had his post-doctoral training in the California University and other Universities of U.S.A.

Dr. Barooah travelled widely various parts of the world, such as U.S.A., Canada, U.K., Germany, Italy, France, Holland, Denmark, Mexico, Japan, Thailand, U.S.S.R. under Rockefeller Foundation Fellowship, Colombo Plan, F.A.O., other organisations and as a delegate to various international meeting representing Government of India. He has travelled very widely in India, visited every State studying Agricultural Development and discussing with officials.

He was a lecturer in the Allahabad Institute, Allahabad University and then worked in various capacities in Assam as, Economic Botanist, Deputy Director Agriculture, Special Officer Food Production, Joint Director of Agriculture, Additional Director of Agriculture, Director of Fisheries, Deputy Secretary (Food Production) and as Director of Agriculture.

In the Government of India Dr. Barooah worked as the Director of Operation in the Central Tractor Organisation, Director of Extension Training, Joint Commissioner (Commercial Crops) and Fertiliser Commissioner. He is the Founder Vice-Chancellor of the Assam Agricultural University. He was the Executive Director of the Pesticides Association of India.

Dr. Barooah is now the working chairman of the State Level Committee of Pathar Parichalana Samity of Assam Government, President, Socio-Economic Development Society for Assam and President, Federation of Agriculture and Allied Services Association.

AGRICULTURAL RESEARCH AND DEVELOPMENT

Role of Scientists and Technocrats

SATYA RANJAN BAROOAH
Founder Vice-Chancellor
Assam Agricultural University

CONCEPT PUBLISHING COMPANY, NEW DELHI 110059

ISBN 81-7022-479-9

First Published 1993

Published and Printed by
Ashok Kumar Mittal
Concept Publishing Company
A/15-16, Commercial Block, Mohan Garden
NEW DELHI-110059

Lasertypeset by
Printing Express
H-13, Bali Nagar
NEW DELHI-110015

PREFACE

This book is dedicated to the Technocrats, Scientists and Specialists in India who all can be categorised in general term as "Technocrats".

This covers all the technical professions in our country such as Agriculture, Animal Science, Veterinary, Dairy, Horticulture, Fishery Soil Conservation, Farm Machinery, Forestry, Home Science, Engineering, Medical, Science and Technology of all categories, Defence, Police, Railways, Telecommunication, Public Health, Architecture, Education, Computer Science, Library Science, Information etc. and all those who specialise in their own profession.

The Technocrats with their specialised training, hard work, sincerity and dedication to the welfare of the nation should be given due weightage similar to those of the generalist, call 'bureaucrats'. There should be no discrimination.

The book is divided into ten chapters. The first chapter gives the experience of my own as a technocrat covering a period of 55 years – as a student at the Allahabad Agricultural Institute till now when I am engaged in two very big programmes as a technocrat. This chapter tells of the problems I faced during the past 55 years and how I have over come. This also gives the satisfaction of work as a technocrat and of difficulties and frustrations that I faced. These experiences of mine will be more or less true with most of the technocrats in any sphere of work. I was lucky to a great extent.

Chapters 2 and 3 deal with the growth of Agriculture Department, the prevailing service conditions, history of Agricultural Services, formation of Indian Agricultural Services, evaluation and assessment of development work as suggested by the Natagarh Committee's Report – which is known as "The Report of the Agricultural Administrative Committee of 1958".

Chapter 4 is one of the important ones as it deals with the training of personnel and also their assessment. This also has been recommended by the various committees but unfortunately have not been implemented properly. In order to build up the competency, training and assessment is a must.

Chapter 5 deals with budgeting, sanctions, procedure and delegation of powers. It is very essential that the powers are decentralised to make everyone very effective. This has also been recommended by the Committees.

Chapter 6 deals with the Anatomy of the Indian Council of Agricultural Research, Government of India. The I.C.A.R is a huge organisation with a large number of institutes being supported by it. The I.C.A.R. Review Committee and the Committee set up for working out the strategy for Eighth Five-Year Plan have made very valuable recommendations for speedy implementation of the various projects and programmes.

Chapter 7 deals with the views expressed by the three late Prime Ministers on technocrats being made Secretaries to the Government of India. The views of the senior officers of the Indian Council of Agricultural Research and the Ministry of Agriculture are also given. These are very revealing. Every one is for technocrats being the heads of the technical Ministries and Divisions.

Chapter 8 deals with the pay and perks of bureaucrats, training and role of political leaders. It is very interesting as narrated by the various administrators in the *Sunday* magazine. The bureaucrats rule the Government – how?

Chapters 9 and 10 are the main thrust of this book. I have brought it out from the writings of the various political leaders, administrators, technocrats and journalists – how most of them agree that the technocrats have been given a raw deal, neglected in case of pay, perks, housing and power in the past. But unfortunately it is still being done when there is a scientific and technological revolution in the country. Can the technocrats be neglected even now ? This is not in the interest of the country when we are heading towards 21st century. This must be remedied and changed. The technocrats also must gear up with more training in their own profession, then on personnel management and administration in Indian institutes and also abroad to take up the challenge. This is what is projected. I have also projected the views how we technocrats should work very closely with political leaders as a team for the welfare of the nation.

In writing the book I have taken help of the reports of all the Commissions and Committees of the distinguished persons appointed by the Government. I wish we implement the recommendations speedily. The reports are gathering dust. The Chief Librarian of I.C.A.R. Mr. S. Moitra has helped me with all the reports and books so also Mr. M.C. Raghavan, Librarian of Indian Institute of Public Administration. I am grateful to them for all the help and allowing me to use the libraries.

I acknowledge with thanks the writers and publishers of the books and reports whose help I have taken. This has been given in the bibliography.

I am thankful to my wife Mrs. Pramila Pandit Barooah, our daughter Kanchan Deshpande and our son Gautam Pandit Barooah in helping me in writing this book by their persuasion.

I am thankful to all who have helped me in completing this important book for the welfare of the country and technocrats.

SATYA RANJAN BAROOAH

CONTENTS

LIST OF PLATES

1

INTRODUCTION

(i) Objective — Why a Book on Technocrat ?

Technocrat in this country are not getting the rightful place and have not been able to play an effective role for various reasons. In the age of rapid scientific developments and High Technology where Agricultural Science have advanced very much it is very necessary that technically qualified persons plan and implement the various development plans in the country where scientific development play the vital role. As such it is only the Technocrats who are qualified and trained should be given the due rightful place, facilities and freedom to man the various technical departments of the Government in the country. But unfortunately it is not happening in India as in the other developed countries of the world.

The developed countries of the world are advancing very rapidly as it is only the Technocrats who are at the helm of affairs of the Technical Departments. In this age of atomic energy, computer science, bio-technology, biological sciences, agricultural sciences who can understand and things better what is needed for the country's development than a Technocrat or a Scientist.

What had been happening in our country is that as far as agriculture is concerned any one and every one seems to be an expert in agriculture and had been managing the departments of agriculture in the country. Hence, the progress has been very slow. It is only lately the thinking at the political level have changed and we are moving ahead haltingly.

It is true of all the other technical departments of the Government such as Engineering, Health, Soil conservation, Horticulture, Forestry, Fisheries, Irrigation, Public Health, Animal Husbandry, other scientific departments where the scientists or the Technocrats had not been given the responsibility to plan and implement the programmes but generalist had been running the departments, and the Scientists and Technocrats only playing a secondary role. Their advise is heard when suited, otherwise thrown to the winds. This has brought frustration in Scientists and Technocrats and stagnation had been created. It is only lately the things are changing, but yet much more to be done so that we can move at a faster rate.

Country cannot wait and see. The people are getting restless and want things to be done, rather getting lost in the redtape of the files, passing from one desk to the other. Very often good proposals and programmes are lost or twisted by the assistants and the bureaucrats endorse them without going into details in order to avoid responsibility. It may not be true in all cases but it is so is some of the cases. This is why the progress of the country is slow compared to the developed nations.

It is time that we stop this situation and move in a more scientific manner by giving the responsibilities of planning and implementation of the various technical programmes to the Scientists and Technocrats. It is also true that the technocrats will require training in administration and personal management which will broaden their outlook and can take decisions taking all factors into consideration rather than form a narrow view point. This is not difficult and can be organised and are being done now. But this requires further strengthening and should be a continuous process.

I have worked in the State Government and also in the Government of India in various capacities, in the State Department of Agriculture and in the Ministry of Agriculture for about thirty years. I have taken up the courage to write this book on *The Technocrats* based on my experience and also other difficulties, hindrances, frustration and some times encouragement in planning and implementation of the various programmes in Agriculture, in the interest of the country, the farmers, consumers and the economic development of the nation as a whole. I feel that I should share with the younger generation all my experience as a Technocrat, and what important role it can play in the future development of the country in all fields of Science and Technology. That is why this book on *The Technocrats*. It will apply to all other branches of Scientific and Technical Departments who are fully involved in the nation-building programmes. I have discussed with engineers who are working in various engineering departments, doctors who are working in the health departments, scientists who are working in various scientific departments and many others who share my experience and views. I have also discussed with a large number of experts such as Agriculture Engineer, Irrigation Engineer, Public Health Engineer, Computer Scientists, Space Scientists, Atomic Energy Scientists, Telecommunication experts, Telegraphs and Postal experts, Railway experts, Power Engineers, Water resources experts, Environment experts, Fishery experts, Veterinary and Animal Science experts, Sericulture experts, Honee Scientists, Economists, Income-tax Officer, Bio-technologists, Geologists, Educationists, management specialists. They all want that something should be done and highlight the shortcoming of the present system. Everyone wants a change for the better.

(ii) My Early Life

I grew up in a small sub-divisional town, now a district town Sibsagar in Assam in the north-east of India. The early life was uneventful. We had large area of cultivable land around our home as we were on one side of the town so I grew up in the agricultural surroundings. We grew paddy, pulses, mustard, fruits, coconuts and vegetables. I took part in the agricultural operations during my holidays from school and also on Sundays. But it was all traditional agriculture then.

My father late Gunjanan Barooah, an outstanding lawyer of his time, and my mother late Homywati Devi had a personality who was respected and loved by all in our town. They encouraged me in my studies. We were eight brothers and one sister out of which I am the seventh one. Two died in early age. All my elder brothers encouraged in my studies and helped me to go ahead, more particularly Late Dharma Ranjan Barooah, Late Gopi Ranjan Barooah and Late Monu Ranjan Barooah. Ours was a very closely knit joint family with a very rich heritage from my fathers and mother's side.

I completed my school education from the Government High School, Sibsagar, Assam and passed matriculation examination from the Calcutta University in first class with distinction in mathematics and geography. Every year was given prizes of books for standing first or second in the class. I was the only student in the history of the school to get distinction in Geography till my time. After completing my studies in Assam I joined the Agricultural Institute at Naini, Allahabad, U.P. — a pioneer city in education.

(iii) As a Student in Agricultural Institute, Naini, Allahabad

This was the turning point in my life — a challenging one. My father was in a serious condition being old age about 80 years, but he insisted that I should proceed to Allahabad to join the Agricultural Institute because he knew that I may not get admission later on. With his and mother's blessings and with the help of my third elder brother Late Gopi Ranjan Barooah and fourth brother late Manu Ranjan Barooah got admission with difficulty as there was competition from all over India for admission as there were only a few agricultural colleges in the country in 1936.

This college was founded by late Dr. Sam Higginbottom of Babson Park, Florida, U.S.A., who was a missionary of the Presbetarian Church. It was with great sacrifice he established the Institute with a band of well qualified American missionaries in the field of agriculture. They taught us apart from theory, practical agriculture by making us work in the field everyday for two hours from 7 to 9 in the morning. The main beauty was the Americans, highly qualified professors used to work with us in the hot sun in the field for the whole period as such it was a great inspiration for us. We learnt how to work with our own hands, the dignity of labour and above all how to handle in a scientific manner the hoe, plough, cultivator, machinery etc.

My four years as a student of the Institute changed my outlook and way of life. Students came from all over India and there was a keen competition among ourselves. The Institute had a very large farm of over 400 acres where various farm crops and horticultural crops were grown. It had also a very good animal farm with a number of breeds of cattle including Jersay and Holstein cows and bulls and buffaloes.

Apart from being Agricultural Science we participated in many social functions, such as debates, farmers fair and competition of various games. The staff members were also very hospitable, kind and co-operative. They invited us to their homes in batches for tea. I became very friendly with the family of Mr. Mason Vaugh, the Head of the Department of Agriculture Engineering and became a part of the family.

The Institute is located across the river Jamuna little away from the Allahabad city. It was right on the bank of Jamuna and a magnificient site to see in the evening when the sun sets across the river. The

Institute was affiliated to the Allahabad University. I graduated from the University in 1940. Really learnt a lot and enjoyed my student life.

(iv) At the Institute of Agricultural Sciences, Banaras Hindu University

I went home after completing the examinations at Allahabad Agricultural Institute and was looking after the management of cultivated land. The result of B.Sc. (Ag.) was announced and I passed in 2nd division standing 2nd in order of merit. I missed my first division by a few marks. It was another turning point in my life. What should I do? I was offered a job by the Department of Agriculture, Government of Assam. But I was more interested in going for higher studies. I got admission in M.Sc. (Ag.) in the Institute of Agricultural Sciences, Banaras Hindu University, Varanasi. It was due to the greatness and large heartness of my two elder brothers, Late Gopi Ranjan Barooah and Late Monu Ranjan Barooah. I joined the Institute of Agricultural Sciences and started my studies in right earnest. We were 21 students in our M.Sc. class from all over India and they were the selected group on merit. So, competition started from the beginning of our carrier in the Institute. It was a Post-Graduate Institute at that time and later on made into a college of Agriculture.

Banaras Hindu University a Residential University with colleges of arts, science, engineering, metallurgy, medicine, library science, law, music and a women's college. The University is situated just outside the city. It has a large number of hostels with all facilities of extra curricular activities. It was started by the great educationist and political leader Pandit Madan Mohan Malaviya with handsome grants from the then Princely States and other Industrialists. When I was a student there were over five thousand students and our Vice-chancellor was Dr. S. Radhakrishnan who later became the President of India, a great philosopher. Now, more colleges have been added and has became a Central University under the Government of India as it was and is a prestigious University.

Our Institute of Agricultural Sciences had only four post-graduate studies than, but later on it has added the under-graduate studies also. It was a very well equipped Institute with very qualified faculty members. My supervisor for the research work where we had to submit a thesis was Late Dr. R.S. Chaudhury an excellent person who guided me and supervised my writing work even at odd hours like four in the morning. I am really grateful to him who died recently. In 1942 after completing two years course passed my M.Sc. in First class standing second in order of merit. While I was appearing in my final M.Sc. examination I received a telegram informing death of our loving mother, and mostly respected Shrimati Homyawati Devi, I cried for the great loss but then I thought I must not lose heart and go ahead with my examinations with full concentration. I never told any one and wrote home later on. She was a great lady and loved and respected by all whoever came in contact with her.

(v) Teaching in Allahabad Agricultural Institute and Allahabad University

I went to Sibsagar, Assam after completing my M.Sc. and getting the result. The then Governor of undivided Assam including Arunachal Sir Andrew Clow asked me if I would join as a Special Officer with headquarters at Shillong to develop agriculture in North-east Frontier Administration now called Arunachal Pradesh. At that time I wanted to go for further studies leading to Ph.D. degree in Agriculture. He agreed after some persuasion and gave me an excellent certificate for a scholarship offered by Canadian Government to the Government of India. I appeared before the Board but not selected due to communal preference given to some one else. However, I have no regret.

In the meantime I got an offer from my old Allahabad Agricultural Institute to join the faculty of the college and teach both Agronomy and Biology. After discussion with my brothers I accepted the offer and returned to Allahabad. I joined the Agricultural Institute as a member of the teaching staff in July 1942. It was a new experience but enjoyed teaching Agronomy and Biology under Rev. B.M. Pugh and Dr. Wellesley. This opened a new chapter in my life, looking forward to new things, new adventure in life. Senior staff members including the American staff members were very cooperative and guided me in my activities. The students were also very cooperative and enjoyed teaching them both theory and practical. Students loved and respected me and never missed any of my classes.

While I was teaching in the Institute I met Prof. Sri Rajan, Head of the Department of Botany where Agriculture Botany was also a part for allowing me to conduct research work in Plant Physiology both in the laboratory and in the field under his guidance. He readily agreed and asked me to get registered in the University for Doctor of Philosophy in Science. It was done since he recommended my application and started my work in both the places simultaneously on the "Effect of micro-nutrients, more particularly of Boron, Zinc, Manganese and Magnesium on the growth of yield of paddy". The results were very interesting and effected the growth and yield of paddy, more particularly zinc and Boron. The research work was completed in 1944 and the thesis was submitted to the University in April 1945. I was awarded the Degree of Doctor of Philosophy in Science in December 1946 and received a cable from Dr. Edgar Vestal, Professor of the Allahabad Agriculture Institute in Berkeley, California University, California, U.S.A. on the 22nd December, the day I married Smt. Pramila Pandit of Nagpur who was studying M. Sc. in Home Science at Oregon State University, Corvallis Oregon, U.S.A. informing of my result. It was a great day of celebration in our life.

When I was teaching in the Institute and also conducting my research work I was appointed as a member of the Faculty of the Botany Department of the Allahabad University to take practical classes in Agricultural Botany. This was a great challenge. I am grateful to Late Professor Shri Ranjan who later became the Vice-chancellor of the University and died recently.

The life in the Agricultural Institute and the Allahabad University was very pleasant and I enjoyed teaching very much. Along with teaching I have learnt a great deal in human relations, more particularly of youth who came from all parts of our country. I was teaching from first year to fourth year student as at that time only under-graduate classes were conducted. Post-graduate teaching was introduced at a later date.

Indian Dairy Diploma course of two years was also being taught at the same time. This was also a very popular course as the Institute has a very good Dairy and produced butter and cheese apart from chilled milk. The Institute has expanded a great deal.

A Diploma course of two years was also taught on Home Science. It was a very practical course for the girls with necessary theoretical courses. I taught botany to the girls. Here I met two very charming girls who became very good friends. One was Miss Pramila Pandit, a graduate of Nagpur University daughter of Rao Bahadur S.R. Pandit, Bar-at-law who was known all over Maharashtra. He was a member of the Constituent Assembly also. Her family was a respectable and well-known family. The other girl was miss Rajkumari Shanti Singh of the Talukdar family of Bhadri a well known family of Uttar Pradesh. Ultimately, I married Miss Pramila Pandit whom I loved and she was a very progressive girl with full of drive and initiative — more about her later.

During my stay at Allahabad Agricultural Institute as a faculty member I officiated as the Head of the Department of Agronomy.

Things slowly started moving for a brighter future, programme and adventure. I applied to the Assam Government for a Post War Scholarship for higher study in U.S.A. Along with it I wrote to the University of California, Berkeley for admission in the post-graduate school and also for a fellowship of Assistantship. In the meantime I completed my research for Ph.D. work and submitted my thesis to the Allahabad University in April 1945.

(vi) Post-Graduate Training in the University of California, Berkeley, U.S.A. and Other Universities of U.S.A. and Canada

Luck favoured me. I got the admission letter of the California University and also a fellowship from the International House, Berkeley. At the same time Miss Pramila Pandit also got a fellowship from the American Home Science Association for post-graduate study in any of the Home Science colleges in U.S.A. She decided to join the college of Home Science of Oregon State University at Corvallis, Oregon.

We both went home to our respective places and prepared in a hectic way to leave for U.S.A. My elder brother late M.R. Barooah helped me with Rs. 5,000 to go abroad. With the blessings of all at home I came back to Allahabad Agricultural Institute and worked for a few months before I finally left for U.S.A. via Delhi and Karachi (now Pakistan) as ships were going from that port to New York direct. All our American senior colleagues gave us introductory letters to various people who later on helped us. Dr. Sam Higginbothom who were well known in U.S.A. for very excellent activity in India was a great help, as also Mr. and Mrs. Mason Vaugh and Dr. E.F. Veste.

Pramila who arrived Karachi early luckyly got a berth in a Cabin in a Passenger-cum-cargo ship in her maiden voyage. Unluckily I could get a berth in a cargo ship. Mr. Divekar, Manager, Reserve Bank of India at Karachi helped me in completing all the formalities.

Karachi was a beautiful city and we stayed at Mr. Divekar's lovely big house for a few days before sailing for America.

The ship of Pramila started a few hours earlier than my ship. We had only six passengers in the aft in gunners cabin. These were called liberty ships used during the World War II. So our adventure to a new continent and a developed country started for our higher studies with a spirit of adventure, without knowing what we will face there. We felt sad that we had to go in different ships but there was no alternative.

I left the shores of our mother land and felt sad but the idea of advance studies and new adventure in a different continent gave me hope for a brighter future, enthused me to look for it. The journey was long and boring as we were only six passengers and two of them were always sea sick. But the beautiful blue waters of the sea, sea gulls kept me busy in the morning and evening. We faced very rough weather as we left port Said and made all of us sick, the ship was rolling from side to side and we passengers were all frightened. After this the journey was smooth and landed in New York after about a month where as Pramila landed after 21 days i.e. before me. Pramila was there with a friend to receive me.

New York is a lovely city. There were lots to see but we wanted to get to our respective universities. We went by train from New York. It was a pleasant journey and quite different from the train of ours, fast, clean and did not stop in stations for long. We parted company at Corvallis and arrived at Berkeley, California and got hostel accommodation at the International House, Berkeley.

International House is a beautiful place built on a small hill with Spanish architecture. We were over 600 students and scholars both men and women. It was primarily for post-graduate students from all over the world of which about 50 per cent were Americans. So we had a good chance to mix and know each other and about their country. The food was good, and there were lot of social activities. We were looked after very well. My boarding and lodging and fees were free because I was awarded the International House Fellowship.

Berkeley campus of the California University was in one side of the city, yet not very far from the main city of Oakland. The other important city very near was Sanfrancisco a very big city with people from all countries doing business or jobs. The climate was moderate, but we used to get fog in the late afternoon. Our campus had about

10,000 students, and all the other campuses together the University of California had over 1,00,000 students.

I wanted to work under Prof. Hoagland who was a eminent scientist on plant nutrition. When I first met him I found him to be a kind and perfect gentlemen always ready to help. He only told me that my doctorate thesis was sent from the Allahabad University for assessment and he found my work very good, original, exhaustive, and recommended for awarding the degree. I was very happy to learn that such an eminent scientist has recommended my work. He was old, must have been over 80 years. So he directed me to another Professor, Dr. Beneth who was working on micro-nutrients and was authority in this field. I worked with him one Semester and then worked with Prof. Stabins Jr. Prof. Jenkins and Prof. Briggs for the 2nd semester in 1946 in genetics and learnt a great deal by working with them.

In the meantime I was offered a grant by the Wattumul Foundation with generosity of Mrs. Ellen Wattumul of Howai, U.S.A. Simultaneously I received the Post-War Scholarship of Government of Assam through the Indian Embassy at Washington, for which I applied to the Assam Government in 1945 before I left for U.S.A. It was a handsome scholarship for two years study abroad. So we were well of to continue our studies. Pramila also got a handsome grant from the Wattumul Foundation for her M.S. degree in Home Science with major in child development.

During the summer break between two semesters I went to Corvallis, Oregon State to meet Pramila and worked in various fields the way the Americans work. The American students work during their holidays, learn dignity of labour and we earn some dollars for our projected trip in future. It was a great experience which we will never forget and urge our students to work during their holidays. We registered ourselves in an employment bureau. They get requests from various agricultural farms, factories etc. We both worked for a cherry orchardist plucking cherries for some time, then in a Hop firm fixing the hop plants which is required for manufacture of beer. Lastly we worked in a fruit canning factory, sometimes day shifts of eight hours and sometime night shifts. Pramila worked in the conveyer belt picking bad fruits and I worked in many capacities from putting the cans in the boxes, to loading of wagons with cans. We earned one dollar per hour each. So we learned and earned and could take trips to various places before we resumed our studies in the next semester.

We went to Mount Hood in Oregon State, two of us and one girl friend of Pramila. As we wanted to save money we hitch hiked, but had very bad experience once in a truck and then in a pick up van. When I think of those experience, I would not advise any one to try hitch hike with strangers.

During our summer holidays at Corvallis we decided to get married on 22nd December, 1946 which is also my birthday. I wrote to my elder brothers and got their approval. I took approval of my mother before she died to marry a girl from outside Assam. Pramila also got her parents approval and blessings before we got married. We prepared our invitation card in a photograph riding on a canoe in a river in Corvallis, Oregon State.

I returned back to Berkeley to continue my post-doctorate studies. It was a distance of 600 miles from Corvallis to Berkeley and took a Greyhound bus which operate throughout the U.S.A. Those are excellent air-conditioned buses, fixed with adjustable comfortable seats like passenger plane seats and very punctual to the minute. It was an overnight journey and enjoyed it very much with regular stops for snacks, breakfast and lunch at their regular bus stops with clean snack bars. All facilities are provided including toilet. It was in 1946. They must have improved very much now.

In April 1947 I completed my studies at Berkeley in the methodology of micro-nutrient studies and also their effect on growth of wheat and simultaneously the methodology of genetical studies and plant breeding so that I could use in my future work in India and more particularly in Assam which was an under-developed State in India.

In the meantime I received the approval of the Educational Advisor to the Indian Embassy in Washington, U.S.A. to visit some of the important Agricultural Universities in U.S.A. and Canada.

Pramila also completed her studies at Corvallis, Oregon State University and was awarded the Master of Science degree in Home Science major in child development. She passed getting all 'A' grades. She was also offered a fellowship for Doctorate degree by the University. I was very proud of her achievement. But she could not continue as I requested her to accompany me in visiting different universities after our marriage. Now I feel I should have encouraged her to complete the Doctorate degree.

On December 22nd, 1946 we got married at Berkeley, California. It was a very interesting and exciting as we had to follow the

American rules of getting our blood tested, and then our marriage was registered according to the laws of U.S.A. and solomonised by a old friend of ours a Presbeterian Pastor who was in India for a long time, in their home attended by a group of friends. The same evening Indian Students Association at Berkeley of which I was the General Secretary organised a lovely reception at the International House where most of the Indian students stayed. It was a lovely and memorable function. Their present a naval clock still remind us of that day. We went for honeymoon at Santa Barbara, a lovely place full of Spanish architecture. We really enjoyed our honeymoon for a week and returned back to our Universities to complete our studies. Lot of publicity was given about our wedding in the local newspapers.

Pramila joined me at Berkeley after completing her master's degree, and we bought a second hand Lincon Zepher 12-cylinder car. It was not the right choice as we should have bought a smaller car. However, we started off on our trip according to the schedule, approved by the Education Advisor to the Indian Embassy and agreed by the various universities to accommodate us for two weeks at each place so that we could study in our respective fields.

We went down South West along the lovely coastal high way to Los Angles and then to Sandiego in California where we were married again according to vedic rights by Swami Paramhansa who built a Temple and also an Ashram. Swamiji was very kind to us and invited all his devotees and others to the function. We felt happy that were married according to Hindu Vedic rites by a great saint who have been preaching Hindu philosophy to the Westerners more particularly Americans who wanted peace and happiness, and our family members were also very happy.

We drove to North from San-Diego and arrived at *Denver,* Colorado where we stayed for one week and met the faculty members of the Colorado State University in our respective fields and studied the latest technique and methodology in improving agriculture. On the way we saw the Grand Canyon where rocks have different colours, and change with the sun's rays. It was an old historic place and a beautiful place. Our next destination was Utah State Agricultural University at Logan. We went through the Salt Lake city, a city built around the lake a beautiful one. Idaho is known as Potato State as potato is grown extensively and they export most of it to other States. I studied the work on potatoes in the various departments. Then we left for Montana State Agricultural University at

Bozman which is a rocky state and primarily dry farming was practised there. We studied the system they were practising. It was entirely a new experience. Professors and Scientists in all the Universities were very kind and willing to show their work and discuss.

We entered Canada through the Alberta province after leaving Montana State. At that time one could freely enter Canada with valid passports. Since we were going towards north we went on driving till 10 at night as there was day light and after stopping Calghary we arrived Alberta State Agricultural University at Edmanton. It is primarily a wheat growing state and one can see miles and miles of wheat because the farms are large and they use mechanised farming. From Alberta we drove to the Saskachwn Agricultural University at Sankatoon, and then to the Manitaba Agricultural University at Winnipeg. These are all called wheat growing States, with other crops in a smaller scale. They also raise dairy, piggery, poultry and even horses. We stayed in the student Union Guest Houses in each place for 10 days and studied their system for agricultural production and extension and my wife Home Science. It was a new experience. Canadians are very hospitable, looked after us well, showed us round and we had access to all their new experiments as to how they can raise the agricultural production of the States. They export a large quantity of wheat to various countries of the world. We returned back to the U.S.A. to study the various techniques and systems of increasing agricultural production and Home Science. We came to Minisota Agricultural University. Agricultural University is at St. Paul where there was a large group of Indian students doing post-graduate studies. After 10 days of study we were at Wisconsin Agricultural University at Madison and stayed for a week. Our next stop was at the Iowa State Agricultural University at Ames. After 10-day visit to the respective department we went to Ohio State Agricultural University at Colombus. It was same as that of Iowa for 10-day programme. Then we visited the State Agricultural University at Tennese, Knoxville. The same routine continued. Lousian State Agricultural University at Baton Rouge was our next destination with the same programme. It is a sugarcane growing state and highly mechanised. Then we moved to the last state on south-eastern coast the Florida State Agricultural University at Gainsville. After completing our university routine programme we visited at Babson Park, our old founder Principal Dr. and (Mrs.) Higginbottom of Allahabad Agricultural Institute in their Orange Orchard. They were very kind and hospitable. Florida is a citrus growing state. We also went to the famous city of Miami, famous for Hollywood stars and had a nice time.

We started our journey to the north through the Eastern Coast and the first stop was at North Carolina State Agricultural University at Raleigh. It was February and very cold and our car was covered with snow in the morning when we came to take it out. The same programme was followed. Then we moved to Washington D.C. The driving was very risky as the road had ice and car started slipping with chains on the wheel. Washington was also cold but we visited Betts-ville, the Experimental Station of U.S.D.A. We met our Ambassador Late Mr. Asaf Ali and the educational advisor Dr. Rosario and finalised our onward journey towards India. We went to the New York State Agricultural University at Cornell and after spending 10 days came down to the New York city for sight seeing. After selling our car Lincon Zephyer in which we travelled from one end of U.S.A. to the other covering 20 States in U.S.A. and three provinces in Canada we got ready to leave for India via London in the famous ship Queen Mary, a luxurious ship with about 5000 persons on board.

We really enjoyed our over 2 years 3 months stay in U.S.A. taking our training in our respective fields which gave us rich experience for its application in our country where we could utilise it with modifications seeing our local environment. All the state agricultural universities where we visited for 10 days or so prepared a very well planned programme from 8 a.m. to 5 p.m. every day except Saturdays and Sundays. This is because we wrote to them from Berkeley giving our entire programme and we kept up the programme. Our Embassy also wrote to them as I was a State Government scholar. The Americans on the whole were very hospitable, kind, showed us round freely and very friendly. They insisted that we stick to our appointed timing which they were very particular. We both are really very grateful to all of them for our memorable stay in U.S.A. in the formative stage of our life. We still maintain a link and visited again in 1963 under Rockefeller fellowship programme for both of us.

During our stay I received a letter in reply to my letter from late Shri Gopinath Bordoloi then Chief Minister of Assam informing me that an important senior gazetted post is kept reserved for me to join on arrival. With great satisfaction we left U.S.A. leaving our lots of friends and professors and scientists both American and Indian.

(vii) Return to India and Join Agricultural Department, Assam Government

We left New York, a beautiful city with lots of things to see around and learn from the hard working Americans. The entertainment centre at Rockefeller Centre was a beautiful place, so also the shops in the 5th Avenue, the statute of liberty and many other things.

Our journey in the Queen Mary was a memorable one, met all kinds of people, full of entertainment, beautiful Atlantic sea with sea gulls around the ship and fabulous variety of food. The few days passed of very quickly.

We landed at Southhampton the port city of United Kingdom not very far from London. We stayed in London for a week and went to all the sight seeing places – the Tower of London where crown jewels are kept and many kings and queens were beheaded in early times. Visited many museums and saw plays in a number of theaters. London is famous for theaters where some plays go on for years. It is fantastic. There were lots of Indians so we did not feel much difference. We also saw the Shakespear's play 'Merchant of Venice' at Stateford-on-Avon. After seeing many more things we took the P and O ship, the city of Hongkong and sailed for Bombay – India our mother-land with high hope.

Our sea voyage in the city of Hongkong was a pleasant one, passengers being mostly Indians and Englishmen returning to their home and place of work, so we developed lot of friendship. The food served was mixed English and Indian like Madras curry, Bengal curry, Bombay curry which were one and the same only names were different. In this ship all were asked to wear formal dress for dinner — the British way of thinking though sometime it was very hot. We were not very far from the shore as such we could see the land most of the time — beautiful sight. We stopped at Gibraltor, Port Said Aden for fresh water, food and other provisions.

We landed at Bombay where Pramila's relations were waiting to welcome us. It was really a very pleasant surprise and we were very happy to land in Independent India in April 1948. We were very happy to meet our relatives. They were also very happy. After a stay of three days we arrived Nagpur to a great welcome by Pramila's parents, brothers, sisters, brother-in-law, their children and lots of friends of Pramila. Those were memorable days. All were very kind and hospitable as they liked Pramila and along with her me also. It became a lasting friendship and love. Whenever we used to go to Nagpur we always enjoyed our stay.

We left Nagpur and arrived Guwahati Airport on Dakota from Calcutta. The airport was built during the Second World War and it was still under control of Indian Air Force. We were welcomed by my relatives and stayed with my eldest cousin sister. It was really very pleasant to be back in Assam after a long period of absence, right from my college days in 1936.

We went to Shillong from Guwahati which is only 72 miles at height of 4500 ft. a very pleasant hill station which was built by the British as the capital of Assam, now Meghalaya. Assam's capital has been shifted to Dispur at Guwahati temporarily.

Immediately on arrival I met late Gopinath Bordoloi, the first Chief Minister of Assam who wrote to me to come back to Assam on completion of my studies in U.S.A. He telephoned the Minister for Agriculture to issue the appointment letter appointing me as the Economic Botanist to the Government of Assam and asked me to see him. I met the Agriculture Minister late Abdul Matlib Mazumdar who requested his Under Secretary, Agriculture, as there was no full time Secretary, Agriculture then to issue the orders. I met him also and he asked me to come again next day. I went to him but the orders were not ready as they had somebody in their mind who was working in the Department of Agriculture. So the red tape and bureaucracy has started its delaying process in spite of the orders of the Chief Minister. I went next day and nothing moved. I did not want to stay in Shillong unnecessarily as I wanted to join my duties and get going with all my enthusiasm. So out of disgust I met the Chief Minister again and he asked me if I got my orders. I told him the problem so he telephoned again the Minister of Agriculture, and also the Under Secretary, Agriculture, to issue the orders that day itself so that I can go back and join my duties. It worked. When I met the Under Secretary he told me that the orders are being typed and I can take the order if I wait for a while. He became very nice to me, talked about his heavy work and how I should go ahead being an young man fresh from abroad. Ultimately I got my orders, thanked him profusely, as we became friends and then met the Director of Agriculture Late Mr. R.C. Woodford an Englishman and the Additional Director of Agriculture Late Mr. L.K. Handique. Both of them were very kind to me and gave me lot of advice and what to do, how to go about, and develop the entire Agricultural Research work in

Assam, as the Economic Botanist was in charge of all the research activities in the State. This is my first experience of bureaucratic interference in spite of the orders of the Chief Minister. This demoralises one's enthusiasm. I learnt different conditions to persevere and get the work done. Bureaucrats can be stumbling block, as some of them do not bother about technocrats. This post was kept vacant for my return to Assam, yet they delayed in issuing the order. So I experienced my first bureaucratic interference 48 years ago.

Pramila who was anxiously waiting at the Hotel was very happy with the news and we celebrated the day and planned our future course of work including our own family. Next day morning we both met the Chief Minister thanked him very much and took his blessings, so also the Agricultural Minister. They both gave their advice and what they expect from me. Then we left for Guwahati and after staying overnight left for home at Sibsagar by train.

It was really a great day for us to come home after our education in U.S.A. and with a prize job of the Government of Assam. We were received both at the station and at home with great enthusiasm, love and affection. All my brothers, sisters-in-law, brother-in-law, sister and my host of nephews and nieces were there to receive us. But I missed my parents, more particularly mother on this day of happiness where the new bride has come home and wanted their blessings. I felt something missing all over. However, others made it up with their warmth and love. It was a strange place for Pramila, entirely new surroundings, strangers and top of it Sibsagar was then a small subdivisional town as compared to a capital city of Nagpur which was the capital of the Central Provinces. Pramila adjusted quickly to the new environment and won over all by her charm and loving manners. But every one looked at her with a strange look being entirely a new person from a distant land of India, and how I married her. All our relatives and friends and others came to see her.

After staying two days at home I borrowed my brother's car and left for Jorhat for joining my new post. There all the people in the Department of Agriculture was surprised as they did not know the orders have been passed appointing me as the Economic Botanist to the Government of Assam. However, they were all good and welcomed me, and I signed my joining report in June 1948. I took over from Mr. P. Ganguli, Assistant Economic Botanist who was acting since partition in 1947.

So I began my new carrier in life looking forward for developing new varieties for Assam of various crops. Got adjusted to my official routine work from 10 a.m. to 4 p.m., and met all officers and assistants in the complex, as all the agricultural research work was concentrated at Jorhat from a very long time, with sections like Agricultural Chemistry, Entomology and Mycology, sugarcane farm and the Economic Botany section looking after all crops in the State. The Economic Botanist was the seniormost officer in-charge of all the section, as he was the Gazetted Class I officer of the State Department of Agriculture. All officers of the Department looked with certain suspicion and jealousy as I was young, only 28 years and coming straight from U.S.A. after higher education. I was only Ph.D. at that time in the Department in Assam.

I picked up my work quickly and also the accounts. Even I passed my departmental examination in Procedure and Accounts within a year of my joining as was advised by the Director of Agriculture. So I concentrated on my work and planned how to improve and expand. Pramila was a great help to me in my work.

The main agricultural crops of Assam are rice, jute, sugarcane, mustard, pulses, vegetable and fruits. The main commercial crop is tea. Arecanuts and coconuts are also grown. I concentrated on rice improvement and also of the pulses. So we went ahead with setting up of sub-stations for deep water rice at Roha near Nowgong and for pulses at Kokilamukh near Jorhat. Land was acquired and estimates for building and staff were sent to the Director of Agriculture, and then to the Government for sanction. The sanction came late as the red-tapism started its work of delay and the bureaucrats without knowing the problems and urgency started delay in sanctioning. Ultimately sanctions came in a mutilated from, and not what I asked for it. So as a Technocrat I got the second taste of government procedure and the attitude of the bureaucracy. Bureaucrats by and large are for following procedures, rules etc. They do not think that rules and procedures can be used for development work. They are indifferent as such the papers are allowed to remain pending in the office and sanctions are issued late as they do not realise the urgency of the technical departments. A technocrat will realise it soon and see that it is done in time as agricultural season do not wait for any one.

I went ahead with full vigour and enthusiasm and completed part of the work of the buildings before 31st March i.e. end of the financial year. But they were not complete, and I found that if I do not spend the money, it will lapse and may take long time again to get it sanctioned next year. So on the advice of my officers and office the money was drawn by a bank draft and paid to the contractors in April. This is a financial irregularity, but there was no other way and not *mala fide*. It was done in the interest of the work, State and the farmers. However, I got a stricture for it after some years, but I did not bother about it as our work was completed to go ahead with the research work on deep water rice. These kind of problems still exist in the Government as they sanction the projects and funds late mostly towards the end of the financial year though some improvement has taken place. Those officers who did not want to take initiative allowed the funds to lapse and projects got delayed.

I brought life into the department by pushing things through, and won the confidence of my then Minister and Director and others. We bought our own car, a plymouth as my work required touring the whole State. It was a good car and toured not only everywhere in greater Assam but all over India later on. When we were in Jorhat Pramila my wife picked up the Assamese language very fast and started a nursery school under the stilted big bungalow with the children of a criminal tribe known as Chowdang. They all appreciated and she got a plot of land near us for a permanent site from the Deputy Commissioner of Jorhat. She became friendly with the American Baptist Mission hospital and they treated the children when needed.

Our first child a daughter was born on the 16th January, 1950 at the Mission Hospital, Jorhat. We named her Kanchan, a very sweet girl, intelligent and now she is married to Dileep Deshpande, General Manager, I.C.I., India, Indian Explosive Division at Gomia son of Mr. Justice V.S. Deshpande who was Chief Justice of Delhi High Court. Kanchan is teaching in the company's school after completing her M.Sc. in Food and Nutrition from Lady Irwin College, New Delhi. She has got two boys, very sweet and intelligent like her studying at Doon School. We used to take Kanchan in a basket in our car everywhere when we went on tour along with our peon (attendant) Sri-Bolorama very trustworthy and good worker.

In 1950 Grow More Food Campaign was initiated by the Government of India with Late Mr. R.K. Patil, I.C.S. (Retd.) as the Food Commissioner in the Ministry of Agriculture and asked the States to do the same with the Chief Secretary as the Food Commissioner. In Assam it was placed under a separate Minister, Late Omeo Kumar Das and Late Mr. S.L. Mehta, I.C.S., the Chief Secretary as Food

Commissioner and Late Mr. S. Majid as the Additional Director of Agriculture, Food Production and Deputy Secretary, Food Production to the Government of Assam in Agricultural Department, thus streamlining the administrative procedure. This helped in getting proposals sanctioned and implemented quickly as the files moved fast and there was very little red tapism and bureaucrats and their assistants were eliminated from the process. I was asked by the Additional Director if I would like to come to Shillong as a Special Officer for Food Production. Pramila and myself were very happy with the idea and accepted the offer immediately and moved to Shillong the Capital of Assam. We felt sorry to leave Jorhat but we had to look for better opportunity and more challenging work.

My work as Special Officer, Food Production which I joined in 1950 at Shillong was to bring under intensive food production the surplus areas of land under the Tea Estates of Assam. Each tea estate had a large acreage which was not under tea and was lying fallow or under grass etc. I visited most of the tea estates and persuaded them to grow food crops thus increased agricultural production in Assam. A large number of them agreed and taken up mechanised cultivation except a few who thought that we were interfering with the activities. But I had the full support of the revenue department and all the Deputy Commissioners of the districts. It was a very challenging and interesting work where I met all types of Tea Planters both British and Indian. Some of the British Planters behaved badly and were rude, did not even offered a chair to sit down, but soon they realised my authority of the State Government and started cooperating. Our idea was to see that they produce enough to feed their own labour who were given rice at a subsidised rate by the Government and the companies.

Over and above this I was put in charge of giving loans to progressive farmers for increasing agricultural production. We helped a large number of them, though some of them were not genuine who tried to pressurise through politicians and Ministers. But we went by merit after thorough verifications. This is the first time I experienced political interference, but could resist them politely by explaining the actual position.

Political interference started playing its role in my carrier as technocrat. The so-called farmers who wanted government loan at a low rate of interest approached me through Ministers and other politicians for sanctioning the loan without much scrutiny. I politely told the Ministers and other politicians that we shall consider all those cases sympathetically and with speed, but within the norms laid down by the Government. They all agreed and did not ask again except a few. They brought to the notice of the Chief Minister also, but he agreed with me. So got the test of it.

In 1951 I was promoted to officiate as the Additional Director of Agriculture and Deputy Secretary, Food Production in the Agricultural Department as the Director of Agriculture had gone on leave abroad, and the Additional Director officiated in his place. There was lot of heart burning amongst the other senior officers as I was very young and joined only in 1948. But however I was declared seniormost by the Government because I was confirmed as Class I Officer earlier than others. So there was lot of jealousy in the Department which I could win by my work and working with as a team, and did not show off my position and authority. The Grow More Food Production drive was a great success — my first thrust.

In order to expedite the implementation of the Food Production drive the State Government gave the full powers to me as the Additional Director of Agriculture combined with the Deputy Secretary, Food Production in the Agricultural Department where I could go to the Minister of Agriculture directly for orders rather than through another process of scrutiny of the proposals by a separate secretariat staff which delayed matters for months. I could go to the Finance Department directly also and got their concurrence to the proposals. This was not favoured by the bureaucrats as most of them wanted to show their own authority over the Technocrats without knowing very much of the technical aspect of it. The proposals were delayed in sanctioning thus slowing down the progress of implementation. But we got over it and moved very fast in our work. This is the best way of getting things done fast in the interest of the country. I strongly feel that in Technical Departments the Technocrats should be the Secretary of that Department so also Additional Secretary, Joint Secretary, Deputy Secretary, under Secretary for speedy implementation of work. The generalists have many other fields of work and departments for their future promotion and where they can show better results, as such there is no clash of interest, yet the work of the state moves on speedily which people want. The system I have mentioned is the best system for speedy implementation of the projects and programmes for the benefit of the people and the country. In Government of India it has been introduced in some ministries such as Agricultural Research and Education, Ministry of Science and Technology, Department of Bio-technology, Ministry of Telecommunication. It should be done in all technical ministries and departments.

The work moved on very well. We started a massive campaign for composting all garbage and cowdung in each village. The then Agriculture Minister Late Mr. Omeo Kumar Das and then late Mr. Mohendra Mohan Chaudhury who became the Chief Minister later on toured very extensively and made mass campaign. It showed its impact and farmers took to it. The then Chief Minister Late Mr. Bishnuram Medhi also gave his full support both administratively and politically. I was very happy with my work under that set up, there being no interference from the bureaucrats. This was the ideal system. Massive campaign on compost programme was a great success due to political support. Hence political support is very important.

During this period the Colombo Plan Programme was being implemented in which delegations from India were sent to Canada primarily for studying the latest technological development in Developed Commonwealth countries such as Canada, Australia, U.K. and New Zealand. This was decided in a meeting at Colombo in Sri Lanka by the Commonwealth countries and that is why it is known as Colombo Plan. A six-member delegation of agricultural senior officers of the State Governments was selected for this for a tour of four months. Luckily I was selected by Late Mr. Medhi then Chief Minister of Assam in a Cabinet Sub-committee meeting. I was very happy and told my wife immediately after the meeting, who has a mixed feeling as she was expecting then. The clearance from Government of India and Canada ultimately came and we left New Delhi after all formalities in August 1951 for Ottowa, the capital of Canada. Unfortunately I had to leave my wife in the hospital who delivered a still born child, our second baby and also our first daughter Kanchan at Shillong. Under very difficult situation I had to leave for Canada. However, everything went all right and Pramila went to Nagpur to her parent's place.

In Canada the officers-in-charge of the Colombo Plan arranged a very well planned programme to visit the main Agricultural Universities, Departments of Agriculture in the Provinces and the main Research Institutes covering all the eight provinces and showed us various farming methods and the latest technological developments. It was a great experience and we have learnt a great deal for application in our own country. Canadians were very hospitable, frank and helpful. They looked after us very well in all possible ways. I learnt the latest method of research work, extension and their method of agricultural education which is similar to U.S.A. Our High Commissioner and the senior officers also helped us, more particularly by Miss Rukmini Menon, who was very friendly and helpful.

Canada's agriculture varies from east coast to west coast. They grow primarily wheat in the provinces, i.e. Alberta, Saskachuwan, Manitoba and part of Ontaroo provinces, the other provinces grow fruits, corn, tobacco, oilseeds, vegetables, very good dairy farming, cattle raising for beef, piggery and poultry, and also lot of wood in the west. Now they have struck oil in Alberta and has become very rich. They have good deposit of potash which we import from there. We could get a variety of experience which I used on return back to Assam in November 1951. In Canada also the red tapism was very little and I found the Technocrats and Scientists were primarily responsible for speedy implementation of agricultural programmes. That is why they progressed very rapidly. They were greatly influenced by the Americans being their next door neighbour and Scientists and Technocrats used to meet often in various seminars, conference and used to go for holidays. They also used to join some refresher courses according to their need. Travelling from Canada to U.S.A. was very easy as there were very little formalities then.

In 1952 Community Development Programme and National Extension Services were started with the help of the Ford Foundation of U.S.A. who gave very good support both financially and with a large number of experts. The main thrust was on increasing food production in a few selected districts of each State by taking the latest technologies to the farmers and cutting down the red tape in administration thus creating a model for expansion. Dr. Einswinger of Ford Foundation and Mr. S.K. Dey, the Administrator of the Community Development Programme were greatly responsible for the change under the patronage of our first Prime Minister late Pandit Jawaharlal Nehru who supported fully. This was a very ambitious programme. And I was also inducted to it in Assam as an Ex-officio Deputy Director of Community Development Department under the Development Commissioner to coordinate the activities of the Department of Agriculture in the programme. There was a direct link between the various development departments and funds were earmarked for it and had separate staff in each block that we selected. In Assam we operated in the Rangia Block where I used to travel extensively. We tried to develop the community in the Rangia Block as a whole in all spheres. It was a great success as a model. We stayed in the villages along with villagers.

In the centre in the Ministry of Agriculture of which community development was a part of it created a Directorate of Extension to plan, prepare schemes and implement those schemes in the centre as well as in the states by giving funds and guidance. The Directorate had Extension Training Division, Project Division, Machinery Division, Woman's Division, Farm Advisory Division, Information Division and an Administrative Division. Since then a number of changes have taken place, a few divisions have been merged with some of the divisions of the Ministry and a new division of Training and Visit Programme has been established. It was a big Directorate under an Extension Commissioner who had wide powers. It played a major role in ushering improved practices in Agricultural Development and increasing agricultural production. A large number of American Advisors were advising the various divisions provided by the Ford Foundation and the U.S.A. I.D. Now they are regrouped as Divisions of Extension Wing, Field Extension, Farm Information and Administration.

In Assam a number of projects were taken up and worked very well since the administrative procedure was streamlined with the Development Commissioner at the head, and very little of interference from the generalist.

Our first Prime Minister late Pandit Jawaharlal Nehru had a very wider outlook and appreciated good work and encouraged technocrats, Scientists and Professionals. In 1957 when I was also the Deputy Director, Community Development Programme in addition to my own duties as Special Officer, Agriculture Department, Late Pandit Nehru visited Assam and was to address the villagers of the Dimoria Block near Guwahati. We were informed by the Additional Deputy Commissioner, Kamrup that Panditji is very angry due to some reasons in the morning at Tezpur and the Chief Minister Late Mr. Chaliha wanted me to look after him and keep him busy by telling all about our community development programme in Assam and not to leave him. So when late Panditji arrived accompanied by the Chief Minister I was introduced to him and then I kept him busy telling all about our programme at Dimoria and also showing him weaving by looms and also mustard oil extraction with '*ghanis*'. He was very impressed. We offered a very hot cup of tea which he liked. He then met the village children who were at the spot.

It was May and very hot, yet I could pacify his anger and Panditji began asking lot of questions, so the Chief Minister asked me to sit with him in between them and kept on talking till the circuit house at Guwahati which was 20 km. Pandit Nehru was very interested in developmental activities and more particularly in scientific and technological advancements. That is why during his time the National Laboratories throughout the country were established with the help of Dr. S.S. Bhatnagar whom he made the Director General of Council of Scientific and Industrial Research and also Secretary to the Government in Science and Technology. So full political support to the technocrats, sccientists and professionals is requires. This was my first close experience with our first Prime Minister — a great man.

(viii) Short-term in the Central Tractor Organisation, Government of India

In 1954 the Ministry of Agriculture, Government of India asked the Government of Assam for my services for the post of Director of Operation in the Central Tractor organisation (CTO). I joined the organisation in September 1954 at New Delhi. The Central Tractor Organisation had a fleet of over 100 very heavy tractors with 100 H.P. or more for clearance of land to make it fit for cultivation. In the initial years it worked in the *Terai* region of Uttar Pradesh which was full of jungles, wild animals and infested with malaria. It was a great challenge to the Organisation and the men who worked for it. They cleared the jungles and made it fit for cultivation for the refugees who came from West Punjab now Pakistan. They were hard working Punjabis, primarily Sikhs. There were others also from U.P. Most of them were ex-army officers who have struggled in the beginning and now well settled with big farms growing wheat, rice, sugarcane, mustard, fruits and other crops. It has became a rich area.

The U.P. Government have established an agricultural university known as G.P. Pant University of Agriculture and Technology in 1962 under the Land Grant pattern of agricultural universities in U.S.A. The U.S.A.I.D. and Rockefeller Foundation has helped a great deal. Now it is one of the good universities. It has a very large farm also to partially support the university. This Agricultural University has helped the settlers in increasing agricultural production.

The CTO had Indian Administrative Service Officers as Chairman and Vice-Chairman with a compliment of staff to administer the organisation and a host of technocrats mainly engineers, a few agricultural technocrats and others. I was primarily looking after the operational part of it i.e. in which State and area to take up the programme, as such I came in contact with a number of Chief Ministers of smaller states of Madhya Pradesh where we were operating then and looking after the welfare of the labour, i.e. tractor drivers, handymen, mechanic etc. who were in the field. We had a very big base camp at Bairagarh near Bhopal, the capital of present Madhya Pradesh. I enjoyed my work.

It was here that I found first the dominance of bureaucrats who were at the head. They wanted to do in their own way, according to us — the Technocrats it should have been done in a different way. They started dictating us being the heads of the Organisation. It was resented by us, but we had no other option. We tried to persuade them to follow the line which was technically feasible, but could succeed only to a certain extent. I was convinced that if the Chairman and Deputy Chairman would have been Technocrats the Organisation would have worked very well. It slowly went down and operation was restricted. Finding this situation I decided to return back to Assam in 1956. Before returning back to Assam I had the privilege of meeting the Union Minister for Agriculture and also the Minister of State for Agriculture and explained the situation of the Organisation. I suggested some changes in the working to revamp the Organisation by taking many other steps as they asked me how to improve the big organisation that was set up with great effort with definite objectives. But later I found that those suggestions remained as it is and red tape and bureaucrats never allowed to see the light of the day. Later on after a few years I found that the Government decided to close down the organisation. It was very unfortunate as it was doing very good work for the economic development of the country.

(ix) Worked in Various Capacities in Assam Government

On return back to Assam Government after deputation in the Central Government after a few months as Special Officer I was appointed as the Joint Director of Agriculture a post which was newly created to look after Research and Extension over and above my old duties of food production in Tea Estates and giving loan to the progressive farmers for increasing agricultural production. I was quite happy with my work. I went ahead with vigour setting up new dimensions in research and extension with the generous help of Indian Council of Agricultural Research. I found the problem of interference of the bureaucrats and assistants in the Secretariat as the proposals submitted for sanction for quick implementation remained unattended or moved very slowly from the lower division assistants to the Superintendent, then to the Under Secretary, then the Deputy Secretary and Secretary in the Agricultural Department who forwarded with the approval of the Minister, Agriculture to the Finance Department for their sanction. The same procedure was followed in the Finance Department the lower division assistant, Superintendent, Under Secretary, Deputy Secretary and Finally Secretary, and some times to the Minister, Finance. I had to pursue at every stage to expedite the sanction and was going from one officer to other. In the process the proposal was sanctioned in a distorted manner which I had to implement. Hence the implementation got delayed, and could be started towards the end of the financial year. By the time we could select and appoint the staff and got the worked started the financial year which ends on 31st March was over and the funds lapsed. This red tape and administrative indifference slowed down the progress of work in Assam, but it is true in all over India because of the system of generalists sitting in the secretariat who do not know the ABCD of the technical proposals and they had to be explained over and over again. If technocrats are appointed in these positions the things would have moved fast as we did in 1951, with the drive for food production.

This was and is the system in all most all the States of India except a few and that is why the progress is slow. There were few understanding persons, but they were very few. It was very frustrating, but we had to work under those conditions. The bureaucrats did not allow the technocrats to come up as they were afraid that their promotions will be limited. It is not so, but that superiority complex as well as fear complex was there. A number of ways were suggested to change or alter the system but after a long deliberations only a slight change was made. In Assam the senior most Chief Engineer, Public Works Department was made the Secretary of the Department and so also as Deputy Secretary and Under Secretary but not in other departments.

Unfortunately the worst part of it in India is appointing Indian

Administrative Service (IAS) Officers as Director of Agriculture in some States such as Tamil Nadu, Andhra Pradesh, Maharashtra, Haryana and Rajasthan. How can these generalist give a dynamic push to the agricultural production, research and extension without knowing anything. They take about six months to learn and get feel of the departments and after about two years they are transferred to another department. The work suffers, progress is delayed. This system should be abolished as early as possible. Only Technocrats qualified in agricultural science should be appointed Directors of Agriculture after proper selection from the officers of the Department, based on merit and seniority. I do not think there is any dearth of capable officers in the Department of Agriculture of the various states who are capable and can give the leadership.

In 1957 the Director of Agriculture retired and I took over as the Additional Director of Agriculture and the Director of Fisheries also. In the meantime the Government under the pressure of the bureaucrats and in there wisdon abolished the post of Deputy Secretary (Food Production) and created another post of Deputy Secretary in the Secretariat to assist a full time Secretary of Agriculture. This actually reversed the process of quick implementation of the programmes both plan and non-plan. Under plan programmes lots of funds were allotted to the States for various new agricultural programmes. Due to the change of the system, red tapism and indifference on the part of the bureaucrats in the Secretariat both in Agriculture and Finance, the files did not move fast, decisions were delayed as such the plan could not be implemented fully and the funds lapsed. This resulted in getting less funds next year as it was always based on the performance of the earlier years. This dampened the enthusiasm of the officer of the department and farmers who expected a rapid change in the growth of agriculture and their economy.

In spite of all these hurdles we pushed ahead the food production programmes and also built up the small department of fisheries into a good well organised Department with officers in each district to develop the potentiality of fish culture in ponds, beels (marshy areas) and rivers under an overall charge of a Deputy Director of Fisheries with more or less full powers of mine as I wanted them to move fast and produce more fish in Assam since Assamese people like fish very much, and this is one of their main dishes with rice. Those who can afford they eat fish both at lunch time as well as at dinner.

Fortunately I had good support from the Chief Minister late Shri Bimala Prasad Chaliha and then Minister of Agriculture Late Shri Mahendra Mohan Chaudhury who became Chief Minister later on after I left Assam for Centre. Both of them had full confidence on me and allowed me to function within the rules and regulations of the Government. I still remember a small incident when Late Shri Chaudhury walked to our house which was across the road in Stoneyland at Shillong at about 8 a.m. when we were having breakfast outside in the sun in the lawn and joined us for a cup of tea. We were quite surprised. Then as I went to see him off to his house which was only a five-minute walk. He told me very plainly that the future of agriculture in Assam depends on me and my staff as we were permanent officers, and not on them as they were politicians and may be a minister today and may not be tomorrow. As such I should realise my responsibility and do what I think best for the State of Assam, he will only guide when necessary. I was taken aback and I realised my heavy responsibility and the confidence reposed on me at that young age. My regard and respect for him became very much. Those were the real political leaders with wide perspective.

But at the same time there were others who are different, who interfered in most of the day to day activities. I worked with such a minister at a later stage who wanted me to do things against the rules of the Government and if I did not do he threatened me that he will do harm. But later on due to destiny he disappeared from the scene (died). This is what we call the result of bad deed (*Karma*).

In 1959 all of a sudden the Director of Agriculture Mr. S. Majid died of heart attack and I took over as the Director of Agriculture, along with it as the Director of Fisheries. The then Minister of Agriculture who wanted me to do some irregular dealings was not in favour of me, but the then Chief Minister late Bimala Prasad Chaliha who knew my work very well as well as integrity decided that no one except me should be the Director. So I reorganised the Department and put more life into the Department. It was a tremendous work, had to put in long hours but got full support from my officers and staff at all levels. I decentralised the work by giving more powers to the Additional Director of Agriculture, Deputy Directors of Agriculture, District Agricultural Officers and Sub-Divisional Agricultural Officers. This worked very well and produced desired results as things moved on faster without referring everything to me. I very strongly feel that in the centre and in every state the powers should be decentralised for speedy implementation of the various plan and non-plan programmes. I also offered full support to my Minister, sought his advice and guidance which somehow did not happen due to many factors.

During my tenure as Director of Agriculture with the blessings and advice of the then Chief Minister late Shri B.P. Chaliha organised "Field Management Committees" in each block composed of progressive as well as small farmers who cultivated in the same contiguous area so that the new technology of better farming practices can be introduced such as better varieties of rice and other crops, proper management of soil and water, plant protection measures through our village level workers called Gram Sevek and extension officers. It worked very well where our officers took more interest. We organised seminars in each sub-division calling over 100 selected F.M.C. President/Secretaries of all categories for two days where various problems of the farmers were discussed and solved on the spot with the help of our officers. The farmers were very happy and went home with determination to produce more and improve their economy. This news spread like wild fire and it was published in the news papers. Over ten thousands Field Management Committees were functioning in the State including the hill districts. It was a very big extension force for the Department. Later Bimala Prasad Chaliha, Chief Minister, Assam gave his full support to the programme. This political support is very necessary. This showed how political support can bring a new dynamism to a programme. This programme was a great success.

During this period the Rockefeller Foundation took interest in improving our research programme which was primarily concentrated at Jorhat, Titabor, Karimganj and Shillong. They were also interested in helping us to establish an Agricultural University as we had the Agricultural College at Jorhat and the Veterinary college at Khanapara, Gauhati. Dr. Ralph Cummings, Director of the Rockefeller Foundation in India at New Delhi and myself had a number of meetings and he met the Chief Minister and the Agricultural Minister also. He found the possibility very good and invited my wife who was a M.Sc. in Home Science (Child Development) and myself to visit some of the State Agricultural Universities in U.S.A. and also other countries on the way and offered a very handsome fellowship. Government of India and the State Government agreed for a six months tour.

But unfortunately in 1962 China who was very friendly with us, declared war against India and attacked us on the North-eastern Border. It was a great blow to India who was not prepared, yet our army and airforce fought hard, but had to retreat from the broder under heavy pressure of the Chinese Army who were preparing sometime and yet professed friendship with India. Late Dr. Ram Subhag Singh, Union Minister of State came on a visit to Assam and we were at Kaziranga Game sanctuary when we heard on the radio that Twang an important border town has been caputred by the Chinese. He immediately returned to Delhi and I was asked to take immediate steps to grow more vegetables, more particularly onion and potatoes as that was the season and he feared that Pakistan may cut off the small link between North-eastern States including Assam and the rest of India, near Bihar and West Bengal Border. Potatoes and onions come largely from Bihar and U.P.

After seeing off the Union Minister at Gauhati Airport I returned back to Shillong and prepared a scheme for Rs. 14 lakhs for growing of vegetable which included distribution of seeds, publicity etc., more particularly on the norther bank of Brahmaputra in the Districts of Darrong, Sonitpur (the Darrong only) and Lakhimpur where the army was in large numbers. I discussed with the Agriculture Minister and the Chief Minister who agreed to the emergency proposal and asked me to present it before the Cabinet meeting which was held everyday. After a few minutes discussion it was approved on principle, yet the then Chief Scretary, an Indian Civil Service Officer asked me to get it vetted by Finance. I told him in the cabinet meeting where the Finance Minister was present that we cannot blow hot and cold from the same nostril. Either I should be allowed to go ahead or let it take its own red tape course to be examined by the assistants and bureaucrats. The Chief Minister asked me to go ahead as I was proceeding to the forward areas to organise next day.

I arrived at Tezpur at 9 at night and mobilised all available seeds of potatoes, onions, tomatoes and other vegetables from the trades till 2 a.m. in the morning, organised publicity with publicity vans, instructed the officers to go ahead and implement the vegetable production progamme. I returned back to Shillong at 8 in the morning driving whole night and submitted the scheme formally to the Government for formal sanction.

Chinese, after taking Bomdilla declared cease fire unilaterally otherwise Assam would have been occupied by the Chinese. The

Prime Minister Late Pandit Nehru announced on the radio his sympathy to the people of Assam thinking the Chinese will be in full control of Assam soon. Destiny changed the direction and future. However, our programme was implemented with full vigour and produced a record crop of vegetables where tomatoes were sold for only 25 paisa and less per kilogram. A dramatic development which surprised all, even the Central Government as we had to approach the then Union Agriculture Minister Shri C. Subramaniam to talk to the Defence Ministry to take the vegetables for the army, rather than bringing from other states through contractors. The point I want to emphasise here is that the formal sanction of the proposal came only after two months of "Operation Vegetable" due to indifference and lack of understanding of the urgency of the bureaucrats and their assistants. Had the Secretary, Agriculture had been a Technocrat he would have appreciated the urgency of the scheme and sanctioned in at least three days. This is one of the glaring example how bureaucrats function in a technical department even during extreme emergency. I did not give it up as had the full backing of the Chief Minister, though the Agriculture Minister was obstructive due to political reasons.

In another opration called "Operation Swarming Catterpiller" which was destroying the paddy crop of the Goalpara district now divided into four districts of Goalpara, Kokrajhar, Bongaingoan and Dhubri. I took an extreme step with the help of the Deputy Commissioner Mr. J.N. Das a bureaucrats announcing that farmers will not be given any loan and will be fined Rs. 50 per acre if they do not dust their crop with insecticides which we were giving free to control the menance which engulfed all four districts, and both of us supervised the operation by mobilising officers and staff of the agriculture, Revenue, Publicity and Police. I requested the Chief Minister not to interfere if allegations of imposing fine and denying loan is concerned for at least three days and were not to be traced and any message sent. A number of allegations were sent by the local leaders who were banking on loans without doing any work for controlling the pests as some of them will take some commission. The operation was successful, and we controlled the insects and saved the beautiful rice crop. When I returned back to Shillong and reported to the Chief Minister he handed over to me all the telegrams. I requested him to tell the Assam Assembly which was in session that he has warned the Director of Agriculture (that is me) to be careful in future, but along with it he praised our work in saving the crop. So the bureaucrat in the District Administration was good one and as such they have place for their work and future in general administration and can really do good to the country by working together with Technocrats.

We have taken up a number of programmes for increasing Agricultural production and Fisheries development. I was also appointed as the Controller of Weights and Measures over and above my work as Director of Agriculture and Director of Fisheries. The main task was to introduce the new system of weights and measures in weights in metic system like grams, kilograms, quintals and tonnes instead of the old system of *maunds*. I enforced the new system with drive and vigour by raiding markets where they were still using old system. Police also helped us in this work in our raids. The then Superintendent of Police at Guwahati Mr. Ataur Rahman was a friend of ours. So he helped in our campaign. Since the traders were not complying with rules and regulations I put the fear of police raids in them to achieve our objectives.

(x) Rockefeller Foundation Fellowship to Study Organisation of Agricultural Universities and Extension in U.S.A. and other Countries of the World

As the situation in the State settled down after the Chinese war and normalcy returned I requested the State Government to give me six months' leave to go abroad under the Rockefeller Foundation Fellowship as the Foundation has been pressing for taking the tour organised by them. Mr. Ramesh Chandra who was then Secretary, Agriculture helped me in getting the clearance of the State Government and also my leave though there was opposition from the Minister of Agriculture for reasons best known to him, but the Chief Minister late Shri Chaliha approved my leave for going to U.S.A. and other countries.

In March 1963 Pramila and myself left Shillong, Assam (now Meghalaya) after leaving our daughter and son in the hostels of the schools. They were very young so we felt very sad to leave them but they had to continue their studies. We came to New Delhi, arranged for our visas of various countries with the help of Rockefeller Foundation. Dr. Ralph Cummings, Director of the Foundation worked out a very detailed programme for us and gave us lots of hints. I wrote to all the embassies of the countries we were visiting to arrange for showing us round which they have done very well. We left New Delhi by Air India and arrived in Cairo, the capital of U.A.R. (then Egypt) in the early hours and received by a Government official of the U.A.R. So we began our visit to various countries to study the latest technology of Agricultural Development in those countries and the Agricultural Universities to study their set up.

We went to Rome from Cairo and spent a few days at Food and Agricultral Organisation (F.A.O.) studying their activities in Agricultural Education, Extension and other programmes taken up in various countries including ours. Mr. B.R. Sen, the then Director General of F.A.O. nominated both Pramila and myself as observers to the World Food Congress which was being held at Washington D.C., U.S.A. at the same time we would be there. The F.A.O. is a huge organisation with offices and experts all over the developing countries. We then went to West Germany and stayed at Bonn. We were shown round the Agricultural Institutions and Research Institutions and also took us to show West Berlin. We saw the contrast of East Berlin and West Berlin. A vast difference then. We were the guest of the West German Government. Then we went to Denmark and stayed in Copenhagen, the capital city and shown round the various agricultural institutions and their work. They were rich in dairy and dairy products. We also visited the dairy town of Sweden very near to Copenhagen and went by a steamer for the day. Our next destination was Netherlands where we stayed at the Hague and showed us round their intensive work on vegetables and flowers which were generally grown in glass houses and in the fields in spring and summer. They are known for their beautiful flowers which is a large export item to most of the other European countries, U.K. and U.S.A. The main centre of training in tropical agriculture was at Waghanigan where we spent some time. Then we went to Paris and visited the Agricultural Institutions around Paris and their method of Agriculture Development and Research. We also saw the beautiful city of Paris and various cultural programmes and institutions. The last stop was at London in U.K. We were taken to the famous Agricultural Research Station at Rothamstead where various research works were going on in all the field of agriculture over 100 years. We were shown the Research Station of Imperial Chemical Industries (I.C.I.) at Helsmir. They have also set up a big research station in Bangalore in India. We saw London city, Cambridge University etc. and parts of Scotland. After spending two months in

Europe we sailed by the Luxury Liner Queen Elizabeth to New York. It was a hectic trip in Europe trying to see everything of our interest for which we had gone and also the various tourist and cultural activities. The journey of four days by the Luxury Liner Queen Elizabeth was very pleasant and relaxing with plenty of good food and entertainment. Atlantic Ocean was clam and we could relax on the deck after breakfast. It was really a very enjoyable trip. Unfortunately these have vanished now with the large net-work of airlines in the world.

On landing at New York the representative of the Rockefeller Foundation met us with a warm welcome and took us to our hotel very near to the office and right in the midst of the city. New York was not new to us except it had changed a great deal. We went to the office of the Rockefeller Foundation and greeted by Dr. Guy Baird, the Deputy Director who was in India earlier. We finalised our detail programme in U.S.A., Mexico, Howai, Japan, Thailand, Hongkong, Manila and back to Delhi-Assam. We met the top persons of the Rockefeller Foundation over lunch and then an orientation programme about the activities of the Foundation. The details of our programme, what do observe in depth, whom to meet and who will look after us were all worked out and explained so that no time was wasted during our two and half months of stay in U.S.A. They are really perfect in going into minute details including sight-seeing at each place. The same method was followed by each Agricultural University and a detail programme was handed over to us immediately on our arrival at the Airport or at the hotel. So we had a well planned tour in U.S.A. to study the set up of the Land Grant Agricultural Universities, their system, changes from time to time, courses, evaluation system, administration, integrated functioning of teaching, research and extension, research stations, outside the campus and all other extra curricular activities of the universities. We spent a few days in New York for orientation and sight-seeing.

Our first visit to an Agricultural University was Cornell University, at Cornell in New York State. It is one of the best known university in Agriculture, Home Science and other faculties. We were shown round various activities, departments and discussed the system of Agricultural University, its administration, teaching, research and extension with many deans, professors, scientists and President of the University – their functioning. It was a wonderful experience. As it was the first exposure to the Agriculture University with the working experience as Director of Agriculture of Assam and studying for establishment of an Agricultural University in Assam. We went into details of its functioning, got very good experience. We then went to the Niagra Falls on the border of Canada. It was a wonderful sight and the whole place was buzzing with life with lots of visitors.

We returned to Washington D.C. to study the functioning of the Agricultural department of the U.S.A. what we call the Ministry of Agriculture and Rural Reconstruction, the Central Research Station at Belts Ville, near Washington D.C.

In Washington D.C. the First World Food Congress was held at that time which was inaugurated by the late President J.F. Kennedy and addressed by late Dr. S. Radhakrishnan, the then President of India. We had the privilege of attending the congress as observers. Pramila attended the sessions related to her subject and I attended the other sessions and also the plenary sessions. It was held in the Department of State who built all the facilities for such a big conference. We really gained very much from our experience in attending the World Food Conference.

We started our real mission of studying the system of Land Grant Agricultural Universities. Our first stop was at Raliegh, North Carolina to study the system in detail of the North Carolina State Agricultural University. Dr. Ralph Cunning was a faculty member, then the Director of Research before he came to India as the Director of Rockefeller Foundation. He was known to all for the charming personality and outstanding contribution to research. Since he arranged our programme and helped us to get the handsome fellowship he was keen that we see his own Institution. It was also the right choice, as we learnt a lot and got most of the information from the University and we were treated as VIPs.

After spending 10 days at North Carolina State University we flew into Columbia, Missouri State to study in detail the system of Agricultural University, their teaching, research and extension education activities. We spent about 10 days here also as it was contemplated that Missouri State Agricultural University will collaborate with the proposed Assam Agricultural University when established with the help of ICAR, and USAID assisting it and also the World Bank. They were already negotiating with Orissa State Agricultural University at Bhubaneshwar and Bihar State Agricultural University temporarily located at Patna. The idea of the USAID who was helping in setting up of the Agricultural universities was that Missouri

Agricultural University at Columbia will collaborate with Eastern State's Agricultural Universities as the conditions and cropping pattern is similar, and it will be easy for them to handle the things and coordinate among themselves. The faculty members and the University officials went out of their way to show us round the facilities in the University, their system of working and introduce to us most of the key persons, and the probable staff members who are likely to come to Assam in this collaborative programme. We were very much impressed by their work, interest in our state and willingness to assist us in establishing and developing the Agricultural University in Assam. Actually two of their staff members assisted the development of the Agricultural Engineering and Poultry Husbandry in the Assam Agricultural College at Jorhat and stayed for a term of two years each. They did an excellent work under the USAID programme. So the Missouri Agricultural University was familiar with our state and its Agricultural potentialities. We requested the President (Vice-chancellor), the Deans of the Colleges of Agriculture, Animal Science and Home Science to visit our State, which they did. After our fruitful and happy stay we went to the Texas State Agricultural University at Austin.

Texas State is a very big State with very big farmers, each farmer having over 20,000 acres to 50,000 acres. Their cropping pattern was different and raised lots of beef cattle. We were shown round the various departments and farming operation. Where there was irrigation they have taken to intensive cropping. So, we were exposed to a different system of farming. There were lots of Indian students also doing post-graduate work in different subjects. The people were very hospitable and kind. We have learnt a lot in handling different systems of farming. Now Texas State is entirely different with oil boom and lot of industries and space research.

Then we flew to the Iowa State Agricultural University at Ames. This state is in the corn and wheat belt and dairy cattle. We studied their system of teaching, research and extension education. We have found that it is almost the same in all the Agricultural Universities with slight variation here and there. The people were also also very hospitable, kind and ready to show you round and take you to different places. We stayed over a week and then we were invited to see a hybrid seed growing company at Dekalb near Chicago, Illinois State.

The Dekalb Hybreed Seed Company sent their twin engine small plane to take us to their place from Ames, Iowa. It was a new experience of flying in a small plane at low altitude having a good view of the countryside. The company was primarily engaged in breeding and production of Hybreed maize seeds of very high quality and yield. Here we saw how a private company is engaged in intensive research work with very highly qualified scientists. It has opened our eyes how we also should go ahead and encourage the private companies to do research and development work and not depend entirely on the Research Institution of the I.C.A.R. and Universities. Because of this there was keen competition and the farmers got really very good hybreed seeds and increased their production. In India it is now coming up, but not to that extent as it should be. A few companies are doing very good work, whereas others have not taken up at all. The Government should also encourage the private companies to invest more money on research and development and give them good incentives.

After a week at Dekalb and Chicago we flew to Oregon State Agricultural University at Corvallis. Pramila studied at Oregon State Agricultural University at Corvallis and got her M.Sc. degree in Home Science, specialising in child development. She was also offered a fellowship for her doctorate degree (Ph.D). It was a home coming for her and they were very happy to see her. We studied their system of functioning and met most of the faculty members. It was more or less the same system.

We went to Berkeley, California where I did my post-doctorate study and stayed at the International House as their guest where I stayed as a research fellow. It was my home coming to my old university, and old staff members were happy to see me. I discussed their system of working with the faculty members. The main campus for agriculture and Home Science was at Davis. We also went there for three days, and studied at both the places their system. They are engaged in both basic and applied research in all fields of science. We had very pleasant memories of our stay in 1946-47, when I was also the President of the Indian Students Association of the California University at Berkeley. We met some of our old friends. The system of Agricultural University of education was more or less the same as others.

Our next stop was at Los Angels, the City of Hollywood, cinema world. We spent some time in the Riverside Campus of the California University as that was the citrus belt and developed intensive farming. They were working for a long time in citrus development and offered post-graduate studies. After seeing the Disney Land developed by Walt Disney, world of Fantasy we headed for Mexico City in Mexico. One cannot imagine about Disney Land unless one actually see it. One got lost in the world of fantasy and enjoy very much.

Mexico is a different country and it has its culture influence by Spaniards. The Government of Mexico showed us round their agricultural university, research stations and also the International Wheat Research Station at Cymit where Dr. Norman Berlang, the noble laureate founded the dwarf varieties of wheat which revolutioned the wheat production in the world and brought the Green Revolution in India. It is from Mexico that India imported wheat seeds of dwarf varieties of Sonora and Larma-raho which brought the changes in sixties and made us self-sufficient today. Our agricultural scientists have developed since then newer and newer varieties by continuous breeding and also developed disease resistant varieties and new agronomic practices. It is a continuous process and work is going on in the Indian Agricultural Research Institute and Agricultural University, Ludhiana, Hissar and Pantnagar, unaffected by the bureaucrats interference and red tapeism. Mexicans are like us, hospitable, kind and showed us round. Our embassy also helped us. Their food habits are also more or less same with lot of spices more particularly chillies. We saw the famous bull fight also.

We flew into Honolulu in Hawai State which is part of the U.S.A. Hawai is known for sugarcane and pineapples and beautiful beaches. We studied their system of the agricultural university. It is the same pattern. The East-West Centre at Honolulu has been doing excellent work in research studies on various aspects. We met Mrs. Ellen Wattermull, Chairman of the Wattermull Foundation who helped us with fellowship grants to each of us. They have helped lots of Indians. The Director of Sugarcane Research Station was a friend of Dr. Cummings of the Rockefeller Foundation, so helped us to see the Research work going on and also the mechanised way of cultivation, harvesting and processing of sugarcane. Labour being very expensive, they have developed mechanised method of cultivation of both pineapple and sugarcane.

We flew into Tokyo, the nerve centre of Japan. The Japanese Government and the Indian Embassy helped us during our stay for over a week in Tokyo, Osaka and Kobe. Japan is mainly a rice growing country and with small holdings. They have intensive cultivation of rice, use very high dose of fertilizers and also pesticides and get the maximum yield of rice. Their system of teaching, research and extension is different, though to some extent influenced by Americans after World War II. But they have developed their own system. Their life style is different. They have still their old traditions and food habits which we liked. They are very hard working and innovative. They improve the technology taken from other developed countries and today they are a leading industrialised rich nation. It was a great experience. We can learn a great deal from Japan.

In Japan I received a letter from A.D. Pandit, I.C.S. (Retd.) who was then Special Secretary, Ministry of Agriculture both in-charge of I.C.A.R. and Directorate of Extension asking me to come through New Delhi and meet him as he has selected me for the post of Director of Extension Training in the Directorate of Extension. We were thrilled to get the letter and then offer of a key post in the Government of India. We became home-sick after this and wanted to come home and see our two children Kanchan and Gautam whom we left in the hostels at Shillong.

We stopped at Hongkong for three days, studied their agriculture which was very little as it is a commercial city and free port bubling with life. After seeing around and shopping left for Bangkok, Thailand and skipped over Manila of the Philippines.

We stayed in Bangkok for a week and studied their system of agricultural university. They are also influenced by Americans after the World War II. Thailand is a rice growing country with other crops in-between. Their agricultural university which is not very far from Bangkok was sprawling university with all the faculties. We studied their system in detail. Bangkok is a beautiful city with lots of temples of Lord Buddha. Most of the people are Buddhists.

After completing our study for about six months with the generous grant of the Rockefeller Foundation we headed for India and landed at Palam Airport in September. We were received Dr. (Mrs.) Cunnings and Mr. Matre of Assam Oil Company, a very good friend of ours. We gave a full report to Dr. Cunnings about our study and said we have been benefited by the study and will be very useful for our future development of agriculture and more particularly in setting up of the agricultural university in Assam. He was very happy. He was getting regular reports from New York about us.

(xi) Director of Extension Training

I met Mr. A.D. Pandit, Special Secretary, Ministry of Agriculture as asked by him in his letter. He told me that I have been appointed as the Director of Extension Training in the Ministry and to collect my formal order from the Under Secretary and join as early as possible. I was very happy and requested him to give me a few days time to go back to Shillong, Assam take the approval of the State Government. Joined formally and then relinquish my charge, pack our things, collect our children from the hostels also meet my staff in the various districts and then join in New Delhi. He agreed but asked me to join as early as possible as my predecessor, Mr. J.C. Ramchandani was retiring in a day or two. Actually I attended his farewell part to which I was invited. I met the Extension Commissioner and other senior staff in the Directorate during three days of our stay at Delhi.

This was turning point in our career and life as we were coming to the national scene with all India prospective. The news of my appointment had already spread in Assam as it was published in the front page of *Assam Tribune*, a premier daily of Assam with my photograph. So it was a talk of the Government of Assam and more particularly of the Department of Agriculture. Without wasting time we arrived in Shillong picked up our two children — a happy reunion after six months and then met the Chief Minister, Late Shri Chaliha and told him of my appointment at Delhi. He was very reluctant to release me and asked me to rethink, but I told him the next day that in the interest of the State and because of the wider field of activity for me and my wife we should go to Delhi and join my new assignment. He then reluctantly agreed and given his blessings. The Agriculture Minister created some problem and did not want to release me but when the Chief Minister agreed he also reluctantly agreed. So after a number of farewell functions we left for Delhi by train in mid-September 1963.

We arrived Delhi and temporarily stayed at Assam House and jointed my new assignment as the Director of Extension Training. It was a challenging assignment as I was in charge of training of middle level workers and senior staff of the Agricultural Departments of the states through a chain of Gram Sevak Training Centres and Sevika Training Centres in the States and three Extension Education Institutes, plus planning and implementation of the National Policy on Extension Training.

The Division of Extensions Training consisted of Gram Sevak Training Centres and their coordination throughout the country including preparation of syllabus. Gram Sevika Training Centres including other women programme, Extension Education Institutes for training of Teachers and also senior officers of the State Department of Agriculture, Rural Youth Programme for the drop outs. The USAID provided advisors for each activities according to the understanding with the United States Agency for International Development. The advisors attached to me were very good and drawn from State Agricultural University of U.S.A. and also from USDA. We also invited young farmers for exchange programme with U.S.A. 16 young farmers and farm girls were sent from here after due selection. The 4 H club of U.S.A. sent 16 Americans young farm boys and girls.

These programmes were implemented very smoothly and had full cooperations with the State Governments. Later on the Directorate of Extension was transferred to the Ministry of Agriculture from Indian Council of Agricultural Research under the Special Secretary Dr. M.S. Randhawa, I.C.S. and a Scientist. In between was the Extension Commissioner who was a bureaucrat, and also the secretariat to process the programmes and also get their sanctions. This delayed the implementation of the programmes quickly and started feeling the bottlenecks.

I also initiated the programme of Extension Education Wings in the college of Agriculture and Home Science in the Agricultural Universities, so that the students of the colleges both boys and girls could be trained in extension technique and methodology to carry out their work then they go to the field after graduation. We gave 100 per cent grants to the colleges for five years. Most of the agricultural universities and colleges implemented the programmes. But it took over a year to get the programme approved by the bureaucrats of the Ministry of Agriculture and Finance. In the Ministry had there been an agriculture technocrat as Extension Commissioner the programme would have been implemented very fast and speedily. Pushing back and forth the files took long time and also to convince them to the utility of the programme. This programme is operating in 26 Agricultural Universities as a part of the University with the help of the I.C.A.R. Now it is in the Extension Education Department of the Faculty of Agriculture, Animal-husbandry and Veterinary, Home Science etc. It was a great achievement of my career. It required dogged perseverance.

The young farmers exchange programme as a part of the Rural Youth Programme also took some time to get going as the then Finance Minister, Shri Morarji Desai was against it, but at the intervention of the late great Prime Minister Pandit Jawaharlal Nehru we got the approval and organised the exchange programme. Obstructions were placed in-between by the bureaucrats but they were overcome. Had there been Technocrats we could have done much better. These young farmers were received by the President, Prime Minister, Vice-Presidents and the Agricultural Minister. Late Mr. Nehru and so also Late Mrs. Indira Gandhi took lot of interest and used to spend quite some time meeting, them, talking and giving their advise, were very much encouraging and many photographs were taken with them. Their programme created a lot of interest throughout the country. But unfortunately this programme has been abolished now due to bureaucratic indifference. It is very sad.

I initiated with the support of Late J.C. Mathur, I.C.S. then Additional Secretary, another programme called "Farmers Training Programme" with the help of United Nations Development Fund, Food and Agricultural Organisation and United Nations Educational and Scientific Organisation. In the beginning it was a 200 million rupees programme but then due to indifference of the bureaucrats in the Ministry of Finance, the Programme was reduced to 120 million rupees, that too after a struggle of two years when we were about to lose the programme from UNDP. The delay was due to bungling of the bureaucrats. Most of the funds were coming from the International Agencies yet the hurdles were put in its implementations. Why this negative approach of the bureaucrats? This hold up the good programme. Ultimately we launched the programme and was accepted by all the States. Under this programme we started farmers training centres with 100 per cent grants for five years, both for men and women so that the latest technology of agriculture can be taken to the farmers to increase productivity and economy of their own and the country. It was well received by the farmers throughout India. This is a one of my achievements as the Director of Extension Training in the Ministry of Agriculture, Government of India, in spite of many hurdles put by the bureaucrats. This programme was appreciated by all the technocrats throughout the country. However Late Mr. J.C. Mathur, I.C.S. and Mr. B. Sivaraman, I.C.S. in spite of being bureaucrats helped in getting it sanctioned. These two were outstanding personality with broad vision. They were also the ordinary bureaucrats.

There were many components of the programme, which included functional literacy implemented by the Ministry of Education. The idea was to teach them through the words which they use everyday in their day-to-day work. The Farmers Training Centres have become now a non-plan programme being implemented by the State Governments in collaboration with the agricultural universities. Recently there was a proposal that it should be combined together with the Krishi Vigyan Kendras so that at least every district has one such centre. This programme also provides for peripatetic training programme so that farmers, both men and women need not leave their home but trainers will go near their gram panchayats or Pathar Parichalana Samities where they come after their household work in the morning and return in the evening after training. This proved very popular.

Another achievement was the entire middle level training programme was given a big push by persuading all the States to take it up and see that they functioned properly and efficiently. I had to tour 15 to 20 days in a month to gear up the States and their officials. Sanctions were delayed in the States by most of bureaucrats. This was a constant problem to be solved by me from the centre. Fortunately we had very good Union Agricultural Minister like Mr. C. Subramaniam and (Late) Fakhruddin Ali Ahmed who gave me a free hand to implement the programme. I used to inform them informally the progress and once in a month a brief report was sent for their information. The Secretaries in the Ministry of Agriculture was also very helpful and supported me all occasions. Persons like Mr. B. Sivaraman, I.C.S., Mr. M.S. Randhawa, I.C.S. (a Scientist) and Mr. T.P. Singh, I.C.S. were the Secretaries, and I could walk into their room any time for getting my problems solved. Those were rare bureaucrats.

In my time fortunately we had full political support from Late Pandit Nehru and Late Mrs. Indira Gandhi. I came close to them because of the young farm youths exchange programme, as both of them took great interest.

In 1965 Food and Agricultural Organisation of the United Nations organised a Seminar in Bangkok, Thailand on middle level training programme for the South-East Asian countries. The countries represented were India, Pakistan, Nepal, Sri Lanka, Myanmar, Laos, Thailand, Malaysia, the Philippines, Indonesia and other

International Agencies. There were about 50 delegates, each country sending 2 to 3 delegates. India was represented by three of us and I was the leader of the delegation. It was a one-week seminar.

We discussed the present position of the middle level training programme and how to improve it depending on our situation. We came out with many useful recommendations which helped us in our future planning and reshaping our training programme. The F.A.O. took lot of interest in the development of the middle level training programme, as it was found that for any agricultural improvement in any country the middle level workers who dealt with the farmers play a great role, as such they should be trained properly and continuously. This is very urgent and important.

In the year 1967 F.A.O. in collaboration with the United States of Soviet Republic (USSR) a two months' study tour of USSR was organised to study the programme of their middle level training in different provinces and how they coordinated at the central level. The countries that were invited were India, Pakistan, Thailand, Nepal, Sri Lanka, Malaysia and the Philippines. Though I was the Director of Extension Training at the central level looking after all the middle level training programme yet I was not selected by the ministry at first and they wanted to send one bureaucrat who was not interested in the programme. But due to the persuasion of the F.A.O. representative in India I was selected and the other person was dropped. This is how some of the bureaucrats manipulate and go to foreign countries for training leaving Technocrats behind. It is a pity but very often this happens. Who benefited? Only the officer and not the country. This is how it happened. We were a three-member delegation of which I was the leader. Ultimately I became the informal leader of the whole group of the delegation. It was a rich experience for us. We had two Russian interpreters and two Russian officials who accompanied us throughout the tour along with one official of the F.A.O.

We were given an orientation course in Moscow for a week on the set up of middle level training programme in the State farms and collective farms and other training centres. We were looked after very well and paid for our incidental expenses and the rest i.e. transportation, hotel etc. were paid by the USSR Government. We were their guests. We visited the training centres and collective and State farms in Maldevia province, Georgia province and Uzbekisthan Province. The middle level training was imparted in all disciplines, more particularly practical training. Girls were also taking training in agricultural machinery and tractor driving. The centres were very well run and more emphasis was given on practical training. They had all the facilities. The collective farms and state farms had large acreage of land running into 10 to 20 thousand acres. The whole country was dotted with these farms and managed on a scientific basis.

We left USSR with a very good impression and return to India introduce a few of their methods to improve our training. The Ministry of Agriculture, USSR looked after us very well and so also the officers who accompanied us, the interpreters and also the F.A.O. official Mr. Yarbarow of U.S.A.

Other important programmes which I initiated were strengthening of the three Extension Education Institutes, one at Nilokheri, Haryana, one at Anand, Gujarat and the other at Rajindranagar, Hyderabad. These were established with specific purpose of training the teachers of the Gram Sevak and Gram Sevika Training centres, as we have a large number of them in all the States. These Extension Education Institutes have been playing a very important role. I was very happy that we could help them with more funds and staff. It is gratifying that I could play a major role in the establishment of the fourth one at Jorhat in the campus of the Assam Agricultural University on 5th January, 1989.

I could also strengthen the women's programme also as I felt that women play a major role in agricultural production and 50 per cent of our adult population are women. States such as U.P. was doing very well, but then Chief Minister abolished the women's programme. A great tragedy for our country. I wish I could have done more.

I could achieve all these in a short span of five years in spite of bureaucratic indifference due to wholehearted support of my colleagues. I really enjoyed working as Director of Extension Training.

In 1968 I was selected as the First Joint Commissioner for commercial crops, a new post created to raise production of commercial crops like cotton, jute, sugarcane, cashewnut, black pepper, coconut, acrecanut, cocoa, oilseeds and crops etc. It was a challenging work with so many directorates for different crops to be looked after. I liked the work very much, sanctioned many new schemes for development, and made all the States aware of their potentials. Since then this programme has made tremendous progress and increased their productivity with the help of new technology and bringing them to the doors of the farmers and growers. I was in the take off stage when I was requested by the Assam Government through the Governor of Assam to establish the Assam Agricultural University at Jorhat.

(xii) Founder Vice-Chancellor of the Assam Agricultural University

In early 1969 I was asked by the Assam Government particularly the then Chief Minister, Mr. B.P. Chaliha to return to Assam to establish the new Assam Agricultural University. I left my new assignment with great reluctance, but with a feeling of satisfaction.

In March 1969 I relinquished my charge in the Ministry of Agriculture and left for Assam with Pramila and Gautam our son. We had to leave our daughter Kanchan who was then in the final B.Sc. in Home Science in Lady Irwin College. We arrived in Mariani Station near Jorhat by train and were welcomed by the Staff and Students with great affection and enthusiasm and drove to the Campus at Jorhat. We stayed in a staff quarter in the midst of the staff colony till the Vice-Chancellor's residence was built according to the proposed plan. The main idea of staying within the campus was to keep a close watch over everything going on in the campus and supervise the activities taken up and planned including teaching, research, extension and building construction so that the university could grow fast and meet the aspirations of the people of Assam.

Immediately on joining I left for Shillong the capital of Assam to meet Mr. B.K. Nehru, the Governor of Assam who was the Chancellor of the University, Late Mr. B.P. Chaliha, the Chief Minister, the Agriculture Minister and the senior officials concerned with the Assam Agricultural University. I knew all the officials as I was the Director of Agriculture, Assam till 1963, and used to tour Assam once a year as Director of Extension Training.

I had long discussion with the Governor, the Chief Minister and the Minister of Agriculture about handing over the College of Agriculture, College of Veterinary Science and the Research Stations of both the Departments to the University under the Act. The problem started surfacing with this demand as the Departments and more particularly the Agriculture Minister were not keen to hand over all the them since they will lose their patronage in case of appointment and transfer of staff and other developmental activities including construction etc. However, I did not insist immediately till I settled down properly with the required staff.

The Governor agreed to my proposal to the appointment of the key officers of the University to start functioning and orders were issued by me on my return to Jorhat, the main campus. This was flashed all over the newspapers in Assam including my formal appointment in a big way. Thus the Assam Agricultural University was established by me and proceeded to make into a dynamic and progressive one amongst the three Universities in Assam.

The students and staff arranged for both of us a very grand welcome on the traditional style of Assam at Jorhat Campus and so also by the public of Assam at Gauhati on the *Bohag Bihu Day* function. We found that all were happy in the establishment of the Agricultural University of Assam and my taking over as the founder Vice-Chancellor. People expected change from the routine slow way of moving.

We took over the academic aspect from the Gauhati University the college of Veterinary Science at Khanapara, Gauhati and from the Dibrugarh University the College of Agriculture at Jorhat. So we started functioning with those two colleges. The examinations were due and we conducted the examination in time according to the schedule and also announced the results in one month after the examination. This made all of us work hard but was appreciated by all students and the public of Assam as the results of the other two universities were never announced in time and always got delayed, thus deprived the students from post-graduate studies in the Universities like Delhi and other places. This was given wide publicity by the newspapers of the Assam and Calcutta. This brought some jealousy amongst the other two universities, as they were criticised.

In the new academic season from July 1969 we introduced the semester system in the first year and previous year of M.Sc. on the pattern of the other older agricultural universities in India such as Punjab Agricultural University and Pantnagar Agricultural University. We made progress by introducing from the first year without disturbing the other years and ultimately covered all the classes year by year. This was done not to bring an abrupt change as both staff and students had to adjust to the new system. I brought in change in the system of admission also by conducting practical examination for one hour by testing the prospective students in field work whether they have any aptitude or not for real agriculture. We also selected the best students, and also earmarked some seats for the students coming from real rural background with land and who will go for farming after graduation. My intention was to make them real good progressive farmers who can take the latest technology to the rural areas and side by side produce graduates and post-graduates who will go for research, extension, teaching and work for the university, government and other fields like Tea Estates and Banks. This was a new innovation introduced by me.

The USAID were assisting all the agricultural universities in India under an agreement with the Government of India through Indian Council of Agricultural Research. World Bank also assisted the agricultural universities through International Development Bank (IDA) and so also the Rockefeller Foundation of USA. So due to Pramila and my close contact with them during our stay in New Delhi since 1963 as Director of Extension Training in the Ministry of Agriculture, we could attract all these agencies to assist our Assam Agricultural University. They readily agreed and assisted all possible ways. The USAID worked out a collaborative programme with the University of Missouri at Columbia through the efforts of Dr. Ron Pollock, Deputy Chief of the Agricultural Education Division of USAID. Dr. Ron Pollock was closely associated and set up a small office for the Team of the University of Missouri at Jorhat. This helped us a great deal to go ahead with full vigour and determination.

A team of experts came from the State University of Missouri and prepared a 10-year plan for the University from 1970 to 1980 with all the details in close collaboration with faculty members of the University. Plans were prepared for each faculty, like agriculture, veterinary, Basic Science and also for Home Science of which Pramila was the honorary Dean of the proposed College of Home Science. All details were worked out including man-power requirement, funds, equipments, buildings, curriculum for each faculty and all other details. It was really a very time consuming and hard work for about a month for all of us. These are being used today and institutions like of College of Home Science has come up which was opposed by the then Minister of Agriculture in the beginning for reasons known to him.

The work of the University went smoothly with few small problems here and there which was bound to come. I kept my doors open to staff and students and also public to bring their problems so that I could solve them which was in my power without delay. This worked very well and after sometime I had very few problems to tackle as such devoted time to build up the University into a model one.

Appointed with the approval of the Chancellor Mr. J.K. Chaudhary, a renowned architect who has built the Haryana Agricultural University and others to prepare a master plan for the University at Jorhat, so that we can go ahead on a planned manner without any haphazard construction of buildings, roads etc. A master plan was prepared and was appreciated by all. It was prepared on the fashion of the Indian Institute of Technology (IIT) at New Delhi where all the faculties at the main campus could converge on the main administrative building with the Library and auditorium at the Centre. The idea was that even in rains the staff and students need not go outside to come to the Library, Auditorium, Seminar rooms etc. This would facilitate close contact with each other. Similarly it was planned for the campus at Khanapara.

Apart from the Team, Dean's of college of Agriculture, College of Veterinary Science, Professors of Horticulture, Professors of Animal Science and even the President of the University of Missouri, Colombia, USA visited the University. They discussed the development plans and how they can collaborate with us and even started selecting their staff from the faculties to come and assist us in the development of the University as a whole. They took lot of interest and also the USAID and the Rockefeller Foundation. We also worked out as to how many of our faculty members will go every year for higher training, more particularly for post-graduate work in the Missouri State University, so that our quality of faculties are strengthened.

Along with it Pramila went on working hard for development of the proposed College of Home Science with the Wheat Associates of India, under USAID apart from the Missouri State University. That Wheat Associates agreed to start a Bakery Training Centre with their own funds. It was all finalised as the Director of Wheat Associates visited the University a number of times and she also met them at New Delhi. But this had to be abandoned due to red tapism in the Department of Agriculture. Assam Agricultural University lost the opportunity to add a new dimension to the activities and younger generation of Assam both, boys and girls lost the opportunity to learn a new trade for the socio-economic development of their own and the State. It was very unfortunate as to how red tape and bureaucrats put hurdles on the path of progress. Another experience of political and bureaucratic interference in the path of progress of Assam.

We organised the Association of the Agricultural Universities for the first time with Late Dr. M.S. Randhawa as the President and I was the Secretary. We also organised the first convention and workshop of the Association in February 1970 at Punjab Agricultural University, Ludhiana on the invitation of Late Dr. Randhawa. It was a very successful convention where important matters were discussed. The USAID helped us. Now it is being held every year.

Ultimately entire USAID Plan was put into shelves due to Government of India's difference with the USA on certain major policy matters. The USAID withdrew from India. Many of the State Agricultural Universities such as Punjab Agricultural University, Pan Nagar Agricultural University, Haryana Agricultural University, Karnataka Agricultural University, A.P. Agricultural University and others took full advantage of the USAID and Rockefeller Foundation with experts, equipments, books, foreign training etc. as they started early. We could not utilise it with best of our efforts and intentions due to bureaucratic indifferences in the State Government and Central Government. So, our Assam Agricultural University which was poised for a very good start got a set back. We all who worked so hard felt discouraged, more particularly Pramila and myself as we came to Assam leaving our challenging assignments at New Delhi for a period of five years to build up the Assam Agricultural University as the Founder Vice-Chancellor. The proverbial '*Lahe Lahe*' (slowly and slowly) was difficult to change to move fast. That's why one day Mr. B.K. Nehru, Chancellor of the University told me that "you are moving at a speed of a jet plane whereas others, responsible for the development of the university, are still in the bullock-cart stage; you see the difference and so you must also move little slowly". However, Late Mr. B.P. Chaliha, Chief Minister and Mr. B.K. Nehru, the Chancellor supported in our efforts.

On 25th April, 1989 I laid the foundation stone of the New Auditorium which took years to complete. The College of Agriculture was handed over at Jorhat by the then Minister for Agriculture in a colourful function presided over by the Chancellor, Mr. B.K. Nehru. Similarly, the College of Veterinary Science was handover by the then Minister for Animal Husbandry and Veterinary in a similar colourful function presided over by the Chancellor at Khanapara, Gauhati. There I was permitted to openly say in my welcome address by the Chancellor and the then Chief Minister of Assam Late Mr. B.P. Chaliha, that the words *Lahe Lahe* (slowly slowly) will not exist in our University and we must move fast". This was applauded by the staff and students and other distinguished guests.

Things moved slowly, the grants from the Assam Government was not released timely, primarily due to the indifference of the bureaucrats and their assistants in the Department of Agriculture. They also thought including the then Minister for Agriculture and Animal Husbandry that the University was only a Department under them and not an autonomous body. This was not realised by them, but rather found too difficult to realise as such we had to face many hurdles which have been avoided. However, we established the Assam Agricultural University, made progress even to bring all the North-eastern States closer through our University by allotting seats for each State and Union Territory. I met each Chief Minister personally to send their students to our Agricultural University. Mr. B.K. Nehru who was then Governor of all the North-eastern States and Union Territories (now States) appreciated our approach. It was a challenging task and we loved it.

(xiii) Commissioner, Fertilizer Promotion

I resigned from the University and joined as the first Commissioner, Fertilizer Promotion in the Ministry of Agriculture, Government of India in October 1971 as the Union Ministry of Agriculture wanted my service. The then Prime Minister Late Srimati Indira Gandhi, Late Fakhruddin Ali Ahmed, Union Minister of Agriculture, Mr. B.K. Nehru, then Governor of Assam, Mr. T.P. Singh, then Secretary, Ministry of Agriculture, Mr. H.M.S. Burney, then Joint Secretary, later on Governor of Haryana and Pramila helped me in joining the Ministry of Agriculture.

The Union Minister of Agriculture, Government of India was very keen that I join the Ministry as early as possible as the fertilizer consumption had gone down from the growth rate of 40 per cent to eight per cent in 1970.

We settled down in New Delhi as we got the Government accommodation very soon. Pramila joined the Indian Council for Child Welfare as Hon'y Asstt. Secretary. Our daughter Kanchan started working for the Mobile Creche as she has already got her M.Sc. in Food and Nutrition from the Lady Irwin College of Delhi University. Our son Gautam joined B. Com. course in the DAV College of Delhi University. So we got busy with our respective work with challenging fields.

I had to start from a scratch as this was a new assignment created to boost the consumption of fertilisers, mainly, nitrogen, phosphates and potash. I had only a personal assistant, a very good stenographer to start with. The work of looking after fertilizer procurement and distribution was with the Input Division dealing with Fertilizers Plant Protection and Seeds. I became a part of it. But there was a small section dealing with Organic Fertilizers such as green manuring, compost etc. with the Crops Division. This was ultimately transferred to the Input Division under me. Similarly the Soil Testing Programme which was also a part of Crops Division was transferred under me. This came at a later stage after I prepared a scheme for the rest of the period of the Forth Plan Period (1969-70 to 1973-74). A scheme was prepared for promotion of fertilizer consumption in selected districts of various States where there were potentials with a bare minimum of staff to begin with and making provision for giving grant and loan to the States for implementation of the scheme. A model was prepared for them so that they could go ahead.

Provision for 36 new soil testing laboratories and quality control laboratories were made to give each State one or two laboratories depending on the State. We provided for building, staff and equipment and also running expenditure except land which was to be provided by the States. The bureaucrats of the Ministry were helpful, but the bureaucrats and the assistants of the Finance Ministry were not helpful and took long time in sanctioning the scheme. They went on asking for clarification of various points at various times. However, after great persuasion, the scheme was approved by the Planning Commission and the Finance Ministry. This delay due to bureaucratic indifference is holding back the progress of the country. Most of them do not take positive approach to a technical programme due to their ignorance of the technical nature of the scheme, programmes and projects, as such, as far as possible technocrats should be in-charge of technical divisions. The average consumption of NPK in India in 1971 was about 45 kg per hectare, Punjab, Tamil Nadu and Andhra Pradesh only consumed more per hectare. So we planned the fertilizer promotional programme in selected districts which we suggested to the State after studying all the parameters which will make immediate impact. This was done in order to avoid delay in selection of districts by the States and to minimise the political influence in selection. After issuing the sanction letters I started touring the States to see that the State Governments take immediate action to sanction the scheme to the Directors of Agriculture for its implementation. Wherever necessary I met the Chief Minister, Agricultural Minister, Agricultural Production Commissioners, apart from Secretary, Agriculture, Director of Agriculture and the Additional/Joint Directors of Agriculture in charge of implementing the programme. This helped in speedy implementation of the programme with all vigour alongwith the components of the programme like Soil Testing Laboratories and Quality Control Laboratories in the States. Where there were technocrats as Directors of Agriculture it moved fast, but where there were generalists things got delayed. I had to look into this very carefully and had to make frequent visits to the States. It is strange but it is true that some States where bureaucrats were as Directors of Agriculture the programme moved slowly. One year's time, the fertilizer consumption went up and there was shortage of NPK which we had to import more from all over the world. Due to this situation in 1973 the Planning Commission and the Finance Ministry asked it there was necessity of the post of Fertilizer Commissioner. But we convinced them that it is a continuous process and there should be no slackness in our efforts. Firstly we are still consuming less fertilizer per hectare as compared to developed nations, secondly, new fertilizer factories are coming up. Thirdly, use of balanced fertilizers has become very important. Fourthly the development of the soil testing laboratories and quality control laboratories in the State became a must. In view of the arguments they agreed and never questioned again.

Twenty-four mobile Soil Testing Laboratories were ordered before I joined, but the supplier was dragging its feet due to indifference of the D.G.S. and D. officials. With great effort I could get all the vans supplied by the supplier and gave to the States for their use to analyse soil samples on the spot and give their recommendations of what and how much they should use the NPK. These helped the farmers as they did not have to send their soil to the laboratories. This was one of my achievements in spite of bureaucratic indifference and delays.

The other great achievement of mine was the establishment of a Central Fertilizer Quality Control and Training Institute. My officers and I worked not allowing any grass to grow under our feet. I laid the foundation stone in 1972 and saw that it was completed in 1974 and most of the equipment in position and started functioning. Today it is catering to the needs of the neighbouring countries also apart from our own States. It is also used very much as a Training Institute for the Scientists who are looking after the State Laboratories, apart from as a referee laboratory. Late Dr. P.N. Kapoor, Dr. N. Tripathi and Mr. Sohan Lal helped me in building this institute.

I also prepared a scheme for the use of urban waste by mechanical methods as compost for agricultural production as organic manure. Also included in it setting up of Gobar Gas (Bio-gas) Plants by individual farmers who had 4 to 5 animals alongwith night soil going directly to the pit. I had to work very hard to get the scheme sanctioned from the bureaucrats of the Finance Ministry though Planning Commission Member Mr. B. Sivaraman approved it in one day due to personally following it up. Finance Ministry made a thick file with notes and counter-notes, but ultimately sanctioned when I told them that the Late Prime Minister Mrs. Indira Gandhi and Member Planning Commission want to see the file. Can you imagine a file containing 10 pages scheme swelled to 60 pages or more? This is bureaucratic indifference and negative approach. The attitude of the bureaucrat must change and technocrats should be in-charge of technical departments. However, as we were ready we launched the scheme with speed and touring all the States. Sometimes we get frustrated and get into a mood of giving it up and to take its own time. But I was not made for that. Late Mrs. Indira Gandhi, Prime Minister, Late Shri F.A. Ahmed, Union Agriculture Minister and Mr. B. Sivaraman, Member, Planning Commission, earlier Secretary, Agriculture took great personal interest in this programme. So also Late K.D. Malaviya, then Union Minister of Petroleum. Because of the political support we moved very fast.

Unfortunately this programme of Bio-gas was transferred to the Ministry of Energy. The whole Fertilizer division is not moving ahead as bureaucrats of the Ministry of Agriculture have not filled up 5 Senior posts out of seven vacant posts of which four for a long time. Why? Some posts have even been surrendered. It is very unfortunate, when more thrust is required in this field of activity.

In 1972 our daughter Kanchan was married to Mr. Dileep Deshpande, son of Chief Justice (Retd.) Mr. V.S. Deshpande of Delhi High Court. He was an Engineer in the I.C.I. (India) Ltd. They have two bright sons studying at Doon School — Neil and Deep. Dileep is now General Manager, Indian Explosive Division of I.C.I. (India) and Kanchan is teaching.

In 1972 I led a three member delegation on Agricultural inputs for three weeks to USSR to study the system they are following the Friendship Treaty signed in New Delhi by President Breznev of USSR and our Late Prime Minister Mrs. India Gandhi, a historic document under which a number of protocols were signed. We went to study, the soil testing programme, use of fertilizers on the basis of soil testing, use of pesticides and related subjects. They drew up a detail programme with orientation at Moscow in their Research Institute and also met the senior officials of the Ministry of Agriculture including the Minister for Agriculture. We were received very warmly and took very good care. The lady interpreter was very good in English and also very helpful. We saw their testing laboratories and how they recommended to their state and collective farms, how much fertilizers to be used, what pesticides to be used in a scientific manner. We visited the state farms and collective farms in Maldevia, Georgia and Tashkent in Uzbekistan provinces. We also visited Samarkhand the great historical place from where Moghal Emperor Babar came. Their laboratories in each state farms were very well equipped with the latest equipments and had trained technicians and scientists in-charge. They lavishly treated us. People were hospitable and were ready to give us all the informations we wanted. It was really a very fruitful visit and we utilized some of their methods in our system when we came back. The other two delegates of my team were Dr. R.V. Misra, who was then Assistant Commissioner in my division and Dr. S.K. Mukherjee, Head of the Division of Agricultural, Chemistry of IARI. Fortunately no bureaucrats were nominated to the team by the Ministry. They treated us as V.I.P.'s and showed us all we wanted to see.

In 1973 I was invited by the Japan Asia Centre for one month's visit to Japan to study research institutes dealing with fertilizer, Agronomical studies, soil testing, quality control of fertilizers, fertilizer factories and also to meet officials of the Ministry of Agriculture and Fertilizer Industry. I saw a big fertilizer factory and discussed with the Japan Urea Ammonium Sulphate Association. This Association had a counterpart in India as Japan Urea Centre. We used to import a large quantity of urea and other fertilizers from Japan. After orientation at Tokyo about Japanese Agriculture by the Japan Asia Centre I visited Kyoto, Nara and Osaka accompanied by an official of the Association who knew English well. It was a very educative and pleasant trip. They showed me round the institutions, farms and factories met a large number of scientists, senior officials, businessmen and farmers. The fertilizer factory at Toyo near Osaka was producing urea with over 110 per cent capacity utilization. They are very efficient in their work, hard working and very few strikes. They have made remarkable progress in industrial development after the Second World War.

In spite of Industrial Revolution the Japanese people have maintained their old tradition regarding food habits, religion, culture, dress etc. Of course the new generation have taken to Western dress and life introduced by the Americans, but yet they maintain their old traditions of hospitality, food and religion. They entertain very lavishly, of course at companies account. It is a beautiful country and developed the electronic and other industries in a big way. They were disappointed as I was not allowed to spend one full month in Japan as they planned a very exhaustive programme to show me round. I also missed seeing the other islands and areas of Japan. They looked after me very well.

On my way back home I stopped at Manila as my nephew Dr. Jayanta Madhav was there in the Asian Development Bank. I visited International Rice Research Institute at Las Banias near Manila who have been doing pioneering work on rice improvement. They have a very big collection of germplasm of rice from all over the world. They have evolved very good high yielding varieties which are growing in large scale in India and improving on it. In other fields also they have been doing very good work, manned by highly qualified scientists without interference from bureaucrats and others. That is why their progress of work was very good.

On my return home I met my wife, daughter and son who were waiting for me at the Delhi Airport. It was a pleasant home coming. Our daughter was expecting a baby at that time and she actually delivered on the same night at the hospital, a lovely boy. I was very happy of the coincidence of being a grand-father.

(xiv) First Executive Director of the Pesticides Association of India

After my various achievements in the State Government and Government of India and putting more than 23 years of service in both the State and Central Governments and 3 years in the Allahabad Agricultural Institute I decided to ask for retirement two and a half years before my tenure was to complete in the Central Government to join the Pesticides Association of India as the first Executive Director. The Association wanted me to join much before, but the Government of India was reluctant to release me as I initiated a new programme of Bio-gas plants (Gobar gas) and mechanical composing with an outlay of 90 million rupees (Rs. 9 crores) during the Fifth Five Year Plan. Looking at pros and cons of everything I relinquished my job as the Fertilizer Commissioner and joined the next day as the First Executive Director of the Pesticides Association of India on 1st March, 1975. I enjoyed my work in the Government in spite of all the problems faced by me with the bureaucrats, sometimes from politicians in pushing through the programmes speedily.

In 1975 our son, Gautam who graduated in commerce from Delhi University joined the Mahindra and Mahindra Ltd. in the International Tractor Division as a Management Trainee at Bombay. So we two Pramila and myself remained in Delhi. We felt lonely but we were very busy with our work. Pramila was the General Secretary of the Indian Council for Child Welfare an organisation started by Mrs. Indira Gandhi before she became the Prime Minister of India. In 1975 we bought an apartment in Sagar Apartment Complex, 6 Tilak Marg, a very central place near the India Gate. So we moved into our own three bedroom flat. We are happy here with good neighbours and a Society of Flat Owners of which Pramila is President and I am Public Relations and Social Secretary. Unfortunately byelaws has not cooperated with the flat owners. After a struggle of 10 years the Delhi Apartment Owners Act was passed in 1986 but has not been made very effective as yet. So we are suffering.

In 1977 our Son Gautam married his class friend in school Dr. Indirani Barooah, daughter of late Dr. S.K. Barooah of Jorhat, Assam an old friend of ours. He left Mahindra and Mahindra and then joined Escorts Ltd. They have two lovely daughters, Pooja and Varsha studying at Delhi Public School.

I started giving a new dimension and push to the Pesticides Association of India. Brought out a journal called PESTICIDES INFORMATION for the first time and organised a National Seminar on Toxicology in Pesticides first of its kind in the history of Association. I enjoyed my work by creating new things and new direction. Here also in the Government, I found the bureaucrats who dealt with the pesticides industry created lot of problems. Instead of solving them quickly. I had to deal with the officials of the Ministry of Agriculture, Chemicals and Fertilizers (now chemicals), Industrial

Plate 1. Mr. Fakhruddin Ali Ahmad, former President of India and the author, Executive Director of Pesticides Association of India at the A.G.M. of P.A.I.

Development, Finance and others. The Association consisted of multinationals, large pesticides manufacturing and formulation houses and small scale formulations. As such it was a heterogeneous body and there used to be clash of interests. Even among the multinationals there were differences. I found that there were always clash of interest and individual members represented their cases very often against the interest of the other members.

So in 1977 I parted company with them and joined the Pesticides Formulations Association as their First Advisor in New Delhi to represent their interests since their Headquarters in Bombay do not know what is happening in the Central Government. I liked the work as I have to deal with policy matters only and they are nice and the problems are same. Late Mr. Shegoankar, President and Late Mr. Modi, Secretary of P.F.A.I. were very helpful. I also worked as the Director of Research and Development for the Motilal Pesticides India Pvt. Ltd. since 1976-1986. I enjoyed my work and helped the company in their growth. I worked very closely with Late Mr. Shyam Goel.

(xv) Consultant, Asian Development Bank

In 1981 the Asian Development Bank invited me to be consultant for preparing a project of Bangladesh to increase their agricultural production in the districts with the help of fertilizers, primarily with super phosphate which was not produced in that country and had to be imported. However, they had urea, a nitrogenous fertilizer produced by them to a limited extent. The Asian Development Bank was interested in giving a soft loan for superphospate purchase. Ours was a team of three experts, one Economist from Pakistan, the other a Pakistani now settled in Hawaii, USA was soil scientist and myself as Agronomist. It was a rich experience. We stayed in Bangladesh for one month gathered all informations and discussed with the various officials, visited the Research Institutes, the districts where we met the farmers and local officials. We could not get all the reports and data as it was not well organised. We collected lots of data from the International Agencies like USAID and World Bank etc.

Most of the International Agencies had their offices and were giving lots of aids, so also the other Western and Arab countries. USA had a very big office and were giving aid on 100 per cent grant. Bangladesh is a underdeveloped country with 100 million people, rather over populated and very little resources of their own. However, they were doing their best to develop their economy. Our mission was to help them to increase agricultural production by use of more balanced fertilizers, as they were being assisted in development of minor irrigation by various agencies including Asian Development Bank. Our main contact was the Bangladesh Agricultural Development Corporation who were dealing with procurement and supplies of Agricultural input such as seeds, fertilizers, pesticides, irrigation, agricultural equipments etc. Our stay was very pleasant, people were nice and hospitable and also the large group of Indian experts who were working there in various International Agencies.

After completing our studies in Bangladesh we came to Manila in Philippines where the Headquarters of the Asian Development Bank is situated. We started on writing our report and I submitted my part in 2 weeks, then it was compiled and a full report was given. The Bank ultimately sanctioned a loan of 20 million US dollars for the project. The Asian Development Bank is a big financing institution like World Bank assisting all the Asian Nations in their development. They are doing a very good work, have experts in all the fields and very well managed.

Pramila joined me at Manila, rather she arrived earlier as our nephew Dr. Jayanta Madhav was there was a Senior Officer in the Asian Development Bank. He is doing very well and showed us Manila and the summer resort Bagio. Manila has developed tremendously during the last decade, very clean city with all amenities, highly westernised. The people were very hospitable. The country was under martial law for a long time. Now under a democratic set up.

I visited the International Rice Research Institute at Las Banias about 100 km from Manila where they have developed a large number of new varieties of rice some of which have become very popular in India like IR 8 a very high yielding variety and also resistant to some diseases. They have a large collection of gion plasm of rice from all over the world for carrying out various research activities. The Institute is fully equipped and a large number of experts on various fields are working. There are many Indians in the institute holding very high positions. When I visited the Director General of the Institute was Dr. M.S. Swaminathan who was Director General of the Indian Council of Agricultural Research in India and Secretary,

Ministry of Agriculture. There is a scientific temper to work and they are not bothered by bureaucrats, as such as progress was faster. He was the only Scientist who was Secretary, Agriculture in the Government of India.

(xvi) President, Socio-Economic Development Society for Assam in New Delhi

I was very interested in the development of Assam State as it has got good potentialities and resources. Moreover, I was closely linked with the development of agriculture in Assam since 1948. I was elected as the President of the Assam Association, Delhi in 1980, 1981 and 1982. I wanted to spread the culture, heritage, traditions, songs and dance of Assam among he people of Delhi and rest of the world, so that Assam is known better. During my tenure as the President we could named a road as Lokapriya Gopinath Bordoloi Marg in front of the Assam House, New Delhi a permanent land mark. Then Assam MP's also helped us. As a President constituted a Committee called Socio-Economic Development Committee for Assam, New Delhi which can help Assam Government and other organisations by expediting the sanctions of the various proposals of the State Government, Assam Agricultural University, Gauhati University and Dibrugarh University and other organisations. Ultimately after my term was over as President in which remained for three years, a few of us both retired and working officers in the Government of India at New Delhi founded the 'Socio-Economic Development Society for Assam', New Delhi and registered it under the Societies Act of 1860 in 1986. We helped the Assam Government in getting funds from the Central Government for flood relief and also pursued other projects mentioned to us by the Chief Minister and other Ministers. Had there been proper feed back about the pending projects our Society could have done many more things. Unfortunately the bureaucrats in the various departments did not help the Chief Minister's Secretariat to prepare a booklet with the latest information of the pending projects. One booklet that was prepared due to my consistent request did not help us.

However, the Socio-Economic Development Society for Assam in a short span of time organised a National Seminar on 'Development of Gas-based Industries in Assam' at New Delhi in 1989 in collaboration with the O.N.G.C. and Oil India Ltd. Short and few practical recommendations were made. I also got two central projects sanctioned – (1) The Extension Education Institute at Assam Agricultural University, Jorhat costing almost 1 crore and (2) The Farm Machinery Training and Testing Institute at Bishwanath Chariali with an outlay of three crore rupees. But it requires the cooperation of the Assam Government and the other institutions in giving feed back.

I am more particularly interested in the growth and the development of the Assam Agricultural University. Pramila is also interested in it as she was the pioneer in getting the College of Home Science started in the University by preparing a ten years programme with all details i.e. from 1970 to 1980 when we were there.

I would like to devote some time of life in Social Reform in Assam as Assam needs it to get rid to certain old traditions as has been done by Swami Dayanand by organising Arya Samaj who have brought lot of changes to get rid of social evils. This is a challenge in my life which will benefit all. We have already registered a Society in Guwahati called 'Samaj Sanskar Sangha (Social Reforms Society). The promoters of this Society are all very old important members of society of Assam. I am now the President of the Society. We are organising branches in all the districts of Assam. I am also interested in educating young students of the schools, colleges, universities and others in the menace of use of alcohol, drugs and smoking which are slowly spreading in Assam in a big way. We have already started to work with the help of the State Prohibition Council, Lions Club of Assam, National Service Scheme, Directorate of Field Publicity, Government of India, Oil India Ltd. and other organisations. We are now spreading the 'Samaj Sanskar Sangha' to the farmers by making them members. It will be a mass movement. We shall take up Family Planning Programme also. Now I have handed over to the Association of Pensioners of Assam Government.

(xvii) Working Chairman of the State Level Committee of Pathar Parichalana Samity, Assam Government

The Government of Assam decided after great persuasion to revitalise the old Field Management Committees and organise new ones to cover the entire State on a permanent basis for Integrated Development of the Farming System. Mr. D.K. Bora, the then Minister of Agriculture who played a very important part inaugurated the Field Management Committees Programme on 4th of May, 1990 at

Khanapara, Guwahati.

This programme is now known as Scheme for Self-employment of Unemployed Rural Youths through Pathar Parichalana Samities. So far over 10,000 P.P.Ss have been organised. Over 20,000 P.P.Ss will be organised by the end of 1992.

There is great enthusiasm amongst the farmers. All inputs of the Agricultural Department are being channellised through P.P.Ss on a priority basis, so also the Shallow Tube Well Programme. We are approaching all the Development Departments to channellised their field programmes through the P.P.Ss.

This is a revolutionary organisation of the farmers for the socio-economic development of the farming community. It is expected to generate self-employment for about two lakhs of unemployed rural youths in the 8th Plan period and more in the 9th Plan period.

The Assam Government have allotted about three crores of rupees in the 8th Plan period but the Planning Commission is willing to raise the allocation to about 15 crores. If we take the funds that is being channellised from other schemes of the Agricultural Department the yearly funds availability will be about ten crores yearly. All the other programmes like cow rearing, poultry, piggery, duckery, horticulture, fishery, soil conservation, social forestry, sericulture, dairy development, water shed management, weaving, adult education, family planning etc. will be channellised through the P.P.Ss. As such it will be very broad based.

Necessary training programmes on farm machinery in Farm Machinery Training Centres is being arranged for the farm youths and all farmers and farm women's training being carried out by the subject matter specialists in each P.P.Ss by the training team. This will be a continuous programme.

Women farmers are given a special place in the P.P.S. Each P.P.S. must have at least 10 per cent women farmers in their executive committee. This we have provided in the rules of P.P.S.

In order to spread the programme and for proper coordination the State Government have constituted a State Level Committee of P.P.S. with the Minister Agriculture as Chairman and Dr. S.R. Barooah (myself) as the Working Chairman and some official and non-official members. This Committee have met every quarter to review the progress. This Committee will co-opt other members also. This new approach to extension was appreciated very much in the International Conference on Extension at New Delhi held from 6th to 9th April, 1990.

Plate 2. The author receiving the National Extension Fellow Award of 1991 from Sri Rao Birendra Singh former Union Minister of Agriculture, and Mr. Mohan Dharia former Deputy Chairman of the Planning Commission.

At this Conference I was given by the Indian Society of Extension Education the National Extension Fellowship Award for excellent work on transfer of technology. The award was given by Sri Rao Birendra Singh, Union Minister of Civil Supplies and Shri Mohan Dharia, Deputy Chairman of the Planning Commission was also present. Assam Association, Delhi also felicitated me in their Rongali Bihu function in April, 1991. It was a great day for me on both the occasions. Fortunately I am getting full support from the bureaucrats and also the politicians so far. The chief minister Mr. H. Saijaa and Agricultural Miniisters Mrs. Taimier of Assam are giving full support.

(xviii) **President, Federation of Agricultural and Allied Services Association (FAASA)**

As I have mentioned from time to time about technocrats not getting proper place in the administration and many important senior posts being taken up by the generalists. In order to assert our rights and positions in the administrative structure both in the Centre and in the State some of us have founded a Federation called "Federation of Agricultural and Allied Services Association (FAASA) of which I am the President, Dr. O.N. Singh, former Animal Husbandry Commissioner as Vice-President and eight senior officers of the rank of Additional Commissioner and Joint Commissioners as Secretaries and Treasurer. Already a large number of officers and scientists have become members. This organisation will try to help all categories of technocrats and Scientists of Agriculture, Animal Science, Veterinary, Agricultural Engineering etc.

In order to implement the National Agricultural policy all recommendations on Administration and Administrative structure and functioning made by the National Commission on Agriculture and other important committees like the G.V.K. Rao Committee on ICAR and Ram Krishnaya Committee on Extension which have been accepted by the Government should be implemented immediately. Over and above these (a) Ministry of Agriculture alongwith allied department should be declared as a Technical Ministry, and the Departments of Agriculture and Animal Husbandry, Veterinary. Dairy Development, Fisheries, Horticulture, Soil Conservation and other allied Departments in the States should be declared as Technical Departments; (b) These ministries and Departments in the centre and also in the States should be headed by qualified Scientists/Technocrats in these subjects; (c) The same running scale of pay and promotions after every 5 years like I.A.S. officers should be given to Scientists/Technocrats. There should be no difference; (d) The Scientists/Technical officers should be given all facilities to function effectively; (e) Provision of training facilities like I.A.S. Officers should be made at the state level, central level and international to improve their competence; (f) Advance action should be taken to fill up all the vacant posts of Scientists/Technocrats on a time bound basis; (g) For better coordination and efficient administration both at the centre, state, district and sub-divisional levels Krishi Bhawans should be constructed to house all allied departments.

FAASA have initiated cases in Central Administrative Tribunal (CAT).

(xix) **Political Support and Interference**

As Economic Botanist, Special Officer, Joint Director, Additional Director, Director of Agriculture, Director of Fisheries and Deputy Secretary (Food Productions) of Agriculture since 1948, I had the privilege of working under three Chief Ministers who were stalwarts amongst Assam Political leaders. Fortunately, I had always good relation though I was a junior officer in 1948 when I jointed the Government of Assam. Late Gopinath Bordoloi actually called me back from USA after post-doctorate training and kept aside the seniormost post of the Economic Botanist vacant for me and saw to it that I was appointed immediately on my arrival though then Agriculture Minister and the Under Secretary was interested in another officer of the Department. As Economic Botanist I was in full charge of all Research activities of the Department and I had full support of the Minister of Agriculture and the Chief Minister as such I could go ahead with the programmes in full swing except some interference from the bureaucrats who delayed the sanctions and I was hauled up by the public accounts Committee for making payments after the end of the Financial year. It did not bother me as it was in the interest of the State.

In 1950 when I took over as the Special Officer on Food Production which had provision of giving loans to the progressive farmers I faced political interference to a certain extent. Late Mr. Bishnuram

Medhi was the Chief Minister. He was a difficult and rough person but I won him over. Actually he once asked me in a meeting of all Heads of Departments as to why I examined critically the loan applications recommended by him. My reply was that I am ordinarily liberal as such if any one goes to the Minister or the Chief Minister it used to arouse my suspicion, so I was very careful. After that no questions were asked. The then Agriculture Minister also cooperated fully, as at that time I was also the Additional Director of Agriculture, Director of Fisheries and Deputy Secretary (Food Production).

In one occasion the Minister of Agriculture wanted to take over the powers of appointment and transfer of Agricultural demonstrators numbering over 1000. I gladly offered it by withdrawing the powers from the District Agricultural Officers as I delegated the powers to them. One will be surprised that after about six months while we were on tour, the Minister asked me to take back the powers as he found that these demonstrators used to abuse him in front of his followers for transferring them from one place to the other. It is difficult to please all. So, it came back to me and I delegated to the District Agricultural Officers immediately.

The Agriculture Minister, Late Mahindra Mohan Chaudhary who became Chief Minister later on and Late Omeo Kumar Das, Minister of Agriculture, even told me that they are politicians, as such may be Minister today and may not be tomorrow, but I as a permanent Government servant and will always be there till retirement and hence should do my best to accelerate the progress of agriculture. It was a great surprise to me, that they realised the realities of life and work. This showed their greatness. Late Mr. Medhi selected me to go to Canada as a Member of the Colombo Plan delegation for four months in 1951, though there were other two senior officers at that time. This showed his faith and confidence in my ability and work.

In contrast to what happened in the past, the other Agriculture Minister turned out to be interfering type and created lot of problems. He wanted certain machineries to be purchased by the Department though our engineers and myself did not approve of it as we were not sure of the quality. But he was influenced by others who were interested in it for various extraneous reasons. However, I insisted on his written orders since I was the controlling officer but he had no courage to give us the order in writing. We did not buy except six. I kept the Chief Minister informed and he agreed with me.

In another palapable instance he wanted me to suspend an office assistant in my department by drawing up proceedings and dismiss him for very flimsy ground as he was a potential political leader. After consulting my staff and also legal experts I refused to dismiss him. The Minister telephoned every morning as to what I have done. I requested him to give me his orders in writing for which he had no courage as he was in the wrong. He threatened me that he will ruin my career, but that did not deter me from my stand. Ultimately the assistant resigned and joined politics and became an important political leader.

In another case, the Minister wanted me to reappoint a fishery demonstrator though he was arrested on a murder charge, but was later acquitted. The senior officers of the Fishery Department were against it and in departmental proceedings his services was terminated as he was an undesirable person and there were many allegations against him since he was from his community, he insisted me to revise the orders. I did not do it. So such things went on with this Minister, but unfortunately I was appointed as the Director of Extension Training in the Ministry of Agriculture after our study tour abroad and I left Assam for New Delhi though the then Chief Minister did not want me to leave Assam, but I could convince him that I could help Assam by going to Government of India in a key position. I did help Assam Government and helping since then by sanctioning a number of schemes and getting more proposals approved.

Even during the Chinese War in 1962, when the Chinese army was coming down towards Tezpur (now Sonitpur) I initiated a very big programme for growing of vegetables in that district as then Pakistan threatened to cut off our railway life line on the border of Bihar and West Bengal. Even this was opposed due to political rivalry in that district with the Minister. The newspapers announced my big programme of vegetable growing to feed our army who were in that area and also the local people since most of our potatoes, onions and certain other vegetables used to come from Bihar and north Bengal. I got the scheme passed through in an emergency cabinet meeting and went ahead by working day and night. The farmers produced a bumper crop of vegetables so much so that tomatoes were selling at 25 paise a kilogram. It was entirely due to the approval of the cabinet whereas the bureaucrats opposed it by saying that they will examine the scheme and I should wait till I get the sanction. Will agricultural season will wait till I get the sanction or for anyone or the Chinese with all their army will wait for official formalities? Actually the formal sanction was received after two months by that time I completed the entire operation. This was appreciated by the Government of India also. That is why when I joined the Ministry of Agriculture in 1963, I was also appointed as the officer on Special Duty with Delhi Administration to increase the production of vegetables as they were expensive and not available in plenty. I got a scheme sanctioned from the Central Government with an outlay of 14 million rupees due to the effort of Mr. C. Subramaniam who was then the Union Agriculture Minister and Mr. Dharam Vira was the Chief Commissioner. The area of vegetables was increased from 10,000 to 70,000 acres and also production of vegetables.

Political support is very important for implementing the programmes speedily and successfully. I had full support in the reclamation of the Jungle land at Fallangani near Doom Doma in Dibrugarh district with the help of the Central Tractor Organisation of the Ministry of Agriculture, Government of India in 1957 when Late Mr. Bishnuram Medhi was the Chief Minister and Late Mr. Mahendra Mohan Chaudhary was the Agriculture Minister. It was due to their full support that I could get the reclamation programme implemented to rehabilitate some landless people of Assam for which Central Government gave 50 per cent as grant of the cost of operation. A fleet of 10 very heavy caterpillar tractors of over 100 H.P. operated in that area and cleared the jungles in two years time. We could rehabilitate over 1000 families and they were given grant and loan for cultivation and as well as other expenses. They grew beautiful crops as the land was rich and virgin. It was a great achievement of mine as the Additional Director of Agriculture in-charge of Food production. There was no bureaucratic interference also as I was also working as the Deputy Secretary (Food Production) of Assam where the proposals were initiated by technocrats like me, examined and got it approved by the Minister of Agriculture, Finance Department and the Chief Minister personally. It was done quietly without waiting for the file to answer questions of minor importance. We could settle across the table. This kind of arrangement in all the State Government and even in the Central Ministries concerning Technical Ministries will go a long way in the speedy development of the country.

In a similar occasion we were confronted with a very serious pest attack on rice which was getting ready for harvest in the then district of Goalpara, now Dhubri, Kokrajhar and Goalpara bordering then

East Pakistan, now Bangladesh. The insect pest called "Swarming Caterpillar" appeared all over the area as it is a rice growing area. There was hue and cry in the papers and telegrams started pouring in from Deputy Commissioner, progressive farmers and my own officers. After ascertaining the gravity of the situation I directed on telephone and by telegram the neighbouring district Agricultural officers and their field staff to move to that district with vehicles and dusting equipments and also all the available B.H.C. 10 per cent dust in that area and at Gauhati. Asked the pesticides companies to get more from their Head office at Calcutta where they manufactured and formulated. I discussed the matter with the Agriculture Minister Late Mr. M.M. Chaudhary and the Chief Minister Late Mr. B.P. Chaliha who were also very much concerned. The Assam Assembly Session was on and the members of those areas were grilling them to do something. I apprised them the whole situation, actions taken to give freely the insecticide without any charge as I had powers to do so in emergency situations. I also requested them not to disturb me as I was moving to the district immediately with my technical staff at the headquarters.

On arrival at Dhubri I discussed with my field officers and then the Deputy Commissioner, a bureaucrat, but very helpful and understanding. We mobilised all the field forces of Revenue, Agriculture and Publicity departments and gave wide publicity to use the insecticide given free by the Government and if they do not use no loans will be given later on as well as may be fined Rs. 50 per family and recover land revenue from them. The impositions of fines was irregular on our part, but the Deputy Commissioner and myself thought that this way we can mobilise all farmers and get their field dusted and control the pest thus save the crops. It worked like miracle and later we moved from place to place. With Almighty's grace, we could control the pests and destroy them within three days of emergency operation. We kept vigilance where it may occur again, but the ripening process of rice started and the crop was saved. When I returned to Shillong after a week the Chief Minister and the Agriculture Minister gave me a bunch of telegrams and letters, some 600 about our work and some objecting to the announcement of not giving loan and imposing of fine. I explained the emergency situation prevailing in that area, they understood and appreciated our work very much. But in order to satisfy the opposition leaders and M.L.A.'s in the Assembly, the Chief Minister made a statement of the entire operation "Swarming Caterpillar" praising the efforts of the farmers, staff of all the Departments and the Deputy Commissioner and myself in saving the crops from devastation. He also stated that we were warned not to do irregular things in future regarding loans and fines. It is because of the political support and political will we could achiever such a stupendous task without any worry. This is what is wanted in India for speedy progress and speedy implementation of the programmes.

Political support given by the Chief Minister of Punjab Late Mr. Pratap Singh Kairon to the Punjab Agricultural University at Ludhiana during the early years of establishment of the University, and also by their Agricultural Minister and the bureaucrats because of the Ministers the University grew very fast, established quickly with all the necessary buildings, staff and infrastructure. Similar was the case of Haryàna Agriculture University which had the full political support of the then Chief Minister, Mr. Bansi Lal and also the Agricultural Minister and the bureaucrats. These two Agricultural Universities developed very fast and have made a name in India. Funds were no constraint to them, released in time and also transferred all the Research Stations to the University as per the provisions made in the Act and visualised for the development of agriculture through the University. Unfortunately it did not happen in Assam primarily because the Minister of Agriculture for reasons best known to him and wrong advice by the bureaucrats and others did not give full support and also did not transfer the research stations for a long time till it was made a condition by the I.C.A.R. for release of grants. It was a great task to get the funds released from the State Government. One has to make many visits to the Secretariat unnecessarily. V.C.'s and senior officers should be spared from this work, so that they can devote time for real developmental work. I had to make many trips to Shillong meeting from the lowest assistant to the Minister to expediting the release of grants. Unnecessary delays were made at various levels. With all the funds from the I.C.A.R. and the International agencies, the Assam Agricultural University could have done much better. It required support from all quarters, but unfortunately that was not coming and some interference and delays were made. Political support is very necessary for the growth of the University and to discharge its functions properly.

One of the Agriculture Ministees made certain remarks during the course of the address to the staff and the students of the University stating that he knew more agricultural technology than some staff members and graduates working in the field. This created a problem as the students resented the remarks and argued with him later on in the circuit house, which was not appreciated by the Minister. In order to avoid such problems in future, we thought the best way was to request the Ministers to come and see the progress of the University, but not to address the staff and students. This worked well and we got rid off any controversy in future.

I wanted to build up a very disciplined group of staff and students right from the beginning and I saw in the Western countries, more particularly in U.S.A. where we studied. Punctuality in all aspects of life is very important. It took me some time but ultimately all staff and students and field workers attended to the work in time. I used to go round the campus from 6.30 a.m. to see that classes were taken in time at 7 a.m. farm labourers and their supervisory staff come to work in time. Gradually every one realised the value of time and this news spread all over Assam due to newspaper reports.

The first convocation we held in 1970 which was addressed by the Vice-President of India, Late Mr. R. S. Pathak was acclaimed by all, right from the Chancellor, distinguished guests to the press reporters as a very well organised convocation in the history of Assam and everything went on in time like a clock, with police band playing national anthem and other tunes before and after the function which added to the gravity and colour to the function. All members of the faculty, staff and students worked hard to make it a success and felt proud of it. The distinguished guests who were invited to the convocation also co-operated fully by coming in time, so that there was no disturbance later on. So the discipline which I tried to introduce worked tremendously well and surprised my own imagination.

But this kind of punctuality was not the cup of tea for some of the political leaders. When Late Mr. M.M. Choudhry took over as the Chief Minister of Assam in later part of 1970 and for whom I had great regard and respect and always supported me in my work when he was the Agriculture Minister. We requested him to visit the University Campus at Jorhat when he visited Jorhat. He readily agreed to visit the University at 6.30 p.m. I reminded him that all will be waiting for him and he may kindly come right in time. I requested his Private Secretary and the Deputy Commissioner, Jorhat also to remind him and arrange that day's programme. We arranged everything right from printing of welcome address and felicitation on his

Plate 3. The First Convocation of the Assam Agricultural University, Jorhat, Assam. From right Dr. Barooah, V.C.; Mr. B.K. Nehru, Chancellor and Governor of Assam etc; Dr. G.S. Pathak, Vice President of India; Dr. M.N. Bora, Ex-Dean College of Agriculture; and Mrs. Pramila Pandit Barooah, Hony. Dean College of Home Science.

becoming the Chief Minister on behalf of the University, staff, students and ladies club. We all were ready and waited on the gate from 6 p.m. Staff and students lined up the entire route to the *pandal*. But, unfortunately, time went by, messages were coming from Police that he is coming soon, but did not turn up till 8.30 p.m. as he was taken round to different areas in Jorhat by his political followers, though he realised that it was getting late. We had to give up the function and sent words to the Police and the Deputy Commissioner that the Chief Minister may be informed accordingly. I even sent our vehicles and staff to intercept the entourage on the way. I also went out with Pramila and the President of the Student Union to intercept on the way to request him to come some other day. As we were returning from the town, the Chief Minister's car headed towards the University at 10 p.m. It was three and half hours late. We intercepted on the way and told him that all waited from 6 p.m. onwards for him, and since he did not come till 8.30 p.m. the students became restive and abandoned the function with great regret. He apologised for the inconvenience and realised that the sense of punctuality brought in the University was being diluted. He appreciated the fact. Actually it was the fault of some of the political leaders accompanying him who delayed him by taking to place not in the programme. Those leaders feel that we should have waited like the villagers who waited for hours, but did not realise the importance of punctuality in an Institution like the University. Some party leaders felt and commented. Actually the Chief Minister was at the University in the morning alongwith the then Union Minister of Information Mr. I.K. Gujral and had lunch. Then he came in right time. This was given wide publicity in the press and after this incident all came in time to the Assam Agricultural University whenever we invited them to see the progress of the University. So the discipline and punctuality was one key note in a day-to-day function so that students could be good future leaders. Political leaders should show the way and respect the V.C.'s.

As long as I was in Assam working in the various capacities in Government and as the Vice-Chancellor of the Assam Agricultural University, I found that practically none of the political leaders took any interest in Agricultural Development of the State except the Chief Ministers and the Agriculture Ministers of the State. That too, there were some interference from the Agricultural Ministers as I have already stated before. If most of them would have taken some interest more particularly in their constituency, the progress of agriculture would have been faster. This is perhaps true in all the States except some progressive states where political leaders have their own farms like Punjab, Haryana, Tamil Nadu and Andhra Pradesh.

In the Central level, i.e. Government of India, I got full support from the Union Ministers in the implementation of the various programmes. I have been told and can see that things have changed a little bit. In my days till 1975, I could meet the Ministers easily including the Prime Minister of India Late Pt. Jawaharlal Nehru and Late Mrs. Indira Gandhi in connection with the Farm Youth Exchange Programme with USA and India. As a Fertilizer Commissioner I used to check the godowns in the States and where there were large stocks we did not allot pool fertilizers in spite of the Chief Ministers of the State wrote to our Minister. I kept the Minister informed of the position so that he could explain the position when they called on him.

The only interference I experienced was when I was the Director of Extension Training and was implementing the Farm Youth Exchange Programme from 1967-68 many letters were written by the Chief Ministers of the States to the Union Minister for selection of certain candidates. Those were passed on to me and I placed them before the Inter Ministerial Selection Committee. Some of them were rejected during the process of screening. In one case the Union Minister kept the file of the selected candidates who were approved by the Secretary, Agriculture, Minister of State for Agriculture for over four months as he was pressurised to select one young youth from Delhi. I requested his private secretary to allow me to discuss with him and explain the decision of the Committee. When I met him and explained, he agreed with my views, however, it was kept hanging for four months and we received cables from Washington every week for the names for them to find out Farm Families to host them. It was an interesting programme and very useful for dissiminating technological developments as well as they worked on India's and USA's farm life Ambassadors.

The delay caused lot of problem and misunderstanding between us and 4-H Club Organisation of USA. The entire processing was delayed due to the candidate. In order to avoid any political interference and pressure on us we used to complete the process of selection before 1 p.m. get it typed and the signatures of the members of the selection Committee. By afternoon the file with the recommendations sent to the Secretary, Agriculture and he in turn used to send to the Minister of State and I found in most of the cases the Minister of State also used to send the file to the Union Minister for his concurrence same day or next day and by this process we at the official level avoided any undue pressure on us. I found this to be very effective except only one case mentioned above. Because of our quick action and decision we earned the respect and credibility of the States who recommended their names and also of the 4-H Club Organisation of USA. The President of India, Vice-President of India and the Prime Minister Late Pt. Jawaharlal Nehru and Late Mrs. Indira Gandhi took lot of interest and talked to both the group at length and gave briefing to the Indian groups and clarified the points raised by the USA group. This helped us in implementation and expansion of the programme.

Political support and political leaders are must for speedy implementation of the various programmes of socio-economic development. It is not only the support of the Ministers incharge of the Ministry, but also of the Prime Minister in the Union Government and also the Chief Minister in the States. Apart from the Ministers, the support of the Members of Parliament and Members of Legislative Assembly are also very important for implementation of the programmes. How this can be achieved is a big question. A system need to be worked out by which M.P.s and MLA's of the States can be involved in the socio-economic development of the States and the country as a whole. Unfortunately they do not take so much active interest as they are busy with their own activities. The only way is to have periodic meetings by the Chief Minister and other Ministers of development departments alongwith the Ministers and senior officials of the departments will, brief them with full details of the programme that are under implementation. Then request them to help in its implementation in their constituencies. This will have a big impact in the field, and the district departmental official will also be a part of. A sustained persuasion is required to involve them.

In 1951 when I was first promoted as the Additional Director of Agriculture, Director of Fisheries and Deputy Secretary (Food Production) I requested the then Chief Minister Late Bishnuram Medhi to allow me to meet the MP's from Assam for the various developmental programmes, we have taken up for increased production of Agricultural crops and also of fish. I gave them briefs of the schemes, gave as much details as they wanted and requested them to activate the people in their constituencies to implement these programmes. They promised in the presence of the Chief Minister as he was a great task master, but unfortunately very few took interest and we never met again as a group. Socio-political awareness is a must for the MPs and MLAs and other party workers for assisting the State Governments in their developments work for the benefit of the farmers and others. Agitational approach is not going to help us.

During the Assembly session we used to meet sometime the members of the Consultative Committee of the Agricultural Department of Assam. Some MLAs took keen interest, but most of them were indifferent. The Secretariat also did not convene these consultative committee meetings regularly for reasons best known to them. The bureaucrats were by and large not keen on meeting the MLAs, though in the meetings it is we the technocrats had to answer all the questions. I think the dialogue with the MLAs of the State is very important and beneficial for involving in developmental work the peoples representative.

Political interference was not much in my time, except in petty transfers and appointments. They used to come to the Ministers and through them to me. I agreed to the reasonable requests, but not to unreasonable one. I also made it clear to the Ministers that they may kindly keep in mind the rules and regulations of the department and the Government. I did not have very many unreasonable demands. But I have found that the situation has changed now and the demands for appointments and transfers have become too much. This is a bad and undesirable situation. This has to be stopped. The technocrats will have to take a reasonable stand and not be pliable all the time. They should explain to the political leaders properly.

2

GROWTH OF AGRICULTURAL DEPARTMENTS AND THE PREVAILING SERVICE CONDITIONS

Agricultural Departments in most of the States were established between 1876 and 1905. Till the year 1920 these departments largely consisted of a few Deputy Directors of Agriculture, Agricultural Chemist, Economic Botanist, Entomologists, Mycologist and some supporting staff. As the usefulness of plant breeding and genetics to be recognised, post of crops specialists, Additional Economic Botanist was added to the departments. During the thirties, under the inducements offered by liberal grants from the ICAR and the Commodity Committees, a large number of research posts were created on a temporary basis, but the staffing of the education and extension branches remained unchanged. Staff was added as each individual new scheme was sanctioned with the financial assistance from the bodies referred to above, and some of these posts were made permanent later on. Owing to financial stringency in the States, it became difficult to secure any general expansion of the department, unless financial assistance was available from outside. This often led to starving of research on vital food crops like rice, millets and pulses as the meagre state resources were allocated to research on sugarcane, cotton etc. Thus, conditions for planned growth and development of the department of agriculture did not exist, though the Research Schemes of the ICAR and Commodity Committees, apart from enlarging the field of research, rendered very valuable service by providing employment opportunities for gaining useful experience to those who had specialised in agriculture and associated subjects.

The States should periodically undertake an examination of the structure, objectives and policies of the State Departments of

Agriculture with a view to providing a balanced growth of the department and renunciating programmes and policies in consonance with the needs of the Agriculture of the State. A committee consisting of eminent agricultural administrators, scientists and progressive farmers can usefully undertake this examination once in five years or so.

(i) History of Agricultural Services and Some Recommendations of the Royal Commission on Agriculture

Prior to 1920, only Bombay and erstwhile Central Provinces had a regular Provincial Agriculture Service, but as a result of the recommendations of the Public Service Commission also known as the Islington Commission. Such services were constituted in all the States after 1920, party by absorption of special posts, partly by promotion from subordinate services and partly by direct recruitment. In all the provinces the minimum pay of this service was Rs.250 p.m. and the maximum Rs.750 p.m. The strength of this service was 157 in 1928. Below the Provincial Service, there were Subordinate Services with varying rates of pay, for which the qualification required was a degree or a diploma of the Agriculture college. The scales of pay different greatly, but the minimum starting pay of the upper grade was not less than Rs. 60 and the maximum was not more than Rs.300 p.m. The lower grade of the service consisted mostly of those who held the certificate of two years course of an agriculture college, the minimum pay not being less than Rs. 30 and maximum not more than Rs. 180 p.m. The lower grade of subordinates was eliminated in the thirties as by then all the certificate courses were converted to degree courses.

In 1924 the recruitment to the India Agricultural Services ceased, because of the decision was not to make further recruitment for all India services in subjects which were transferred under the Constitutional Reforms of 1919, to the control of the Governors of the Provinces acting with their ministers. In 1924, the sanctioned strength of the service was 157 – 138 being with the State Governments and 19 with the Government of India. By 1928, only 93 officers were performing duty of these 138 posts in the States and Government of India. It was intended that a new superior Provincial Agricultural Service would be established to take the place of Indian Agricultural Service, but by 1930 the idea of constituting a common Superior

Provincial Service was abandoned. A number of States, however, constituted a Class I service a distinct from Class II service.

The Royal Commission on Agriculture in 1928 also made important recommendations with regard to the service conditions, but unfortunately most of these recommendations have not been implemented and many of present problems of law morale and inefficiency would have been avoided if these recommendations had earned more sympathetic consideration.

The Royal Commission considered the Posts of Director of Agriculture and the Principal of an agriculture college of such importance that they scheduled them as selection posts, outside the cadre of Class I officers. They suggested for the Principals of Agricultural Colleges in the scale of pay of Rs. 1500-50-2000 and recommended that the Director of Agriculture should be placed on terms of equality with the Heads of other departments, such as those of the Public Instruction and Forests. The pay scales of Agricultural Department and others have been revised by the 4th Pay Commission. It was recommended that the officers selected for the Directorship should combine administrative capacity with high scientific qualifications and if it happened that the postulated administrative capacity was not forthcoming in the Department of particular province, the local government should in the first instance, turn to the Agriculture Department of another Province and failing that to the Indian Civil Service. Unfortunately, these recommendations were also not implemented and an increasing number of Directorship went to officers of the Administrative Service.

(ii) Observations made by the Indo-American Team

The Joint Indo-American Team on Agricultural Research and Education in their Report (1955) have stated:

> Some of the major difficulties in the way of developing improved technical leadership in the fields of agricultural research, education and related advisory services are due to current policies of personnel management. Considerable attention has been given to the development of Agriculture and Administrative facilities, including colleges and research laboratories in recent years. There is some indication that the enthusiasm for developing improved physical facilities has been accompanied by a lack of adequate for sight in providing for the type of scientific and teaching staff that

is essential if the new facilities are to function effectively.

The problems relating to professional agriculturists and their administration in India are numerous and complex. They must be analysed carefully and definite steps taken to improve the position of agricultural scientists and educationists, if technology in agriculture in the country is to progress.

The direction or supervision of research or educational programmes in agriculture, as in any other field, must be entrusted to individual who have a full understanding of the problems, requirements and potentialities in these technical fields. The head of the research institution must be a stimulating, driving force capable of perceiving the nature of the complex problems, determining the feasibility of the various research approaches, and must fully understand that research in many instances must be slow, laborious and in numerous instance unproductive. Such proper technical leadership will lend stability and provide for continuous progress in research. The same general qualities are required in the education and extension fields. There is a tendency in some States or institutions to place non-technical administrative officers in charge of agricultural research, education and other technical services. This is the move in mediocrity in technology. It is recognised that individuals who have had the varied experience provided by the practice of shifting from one administrative post to another of continuously higher responsibility do acquire a broad background experience that is valuable in enabling them to meet many diverse and complex problems. They become extremely competent in making decision on broad administrative matters. Progress in the technical field of agriculture, however, requires more than ability to iron out administrative wrinkles. It requires the leadership of career professional agriculturists rather than career professional general administrators.

Science and education are the creative and productive forces for India's agriculture. They must be provided with the highest calibre of technical guidance available in the country. This relates to the key positions of supervision not only in research and educational institutions, but also in the position of state directors of agriculture.

The late Prime Minister Pt. Nehru remarked:

"In order to get the best results, the top positions in Agriculture Ministry and the Agriculture Department in the States should be manned by the technical men with practical experience of agricultural problems and these men should be allowed to operate on their budget allotments. . . ."

(iii) Need for Better Service Conditions

Technical departments such as the Department of Agriculture, Animal Husbandry etc. have not be developed to the same extent in the past as the general administrative services. The index of the development in the country, in this scientific age, specially in the welfare state, is determined by the status and efficiency of the technical staff and their number. If this barometer of progress is accepted, as it has to be, then it does not speak well of the progress made in agriculture, which constitutes the most important sector in our economy.

It is often said that the technical services have not been able to deliver the goods. May be this is partially true. The fault is not so much of the technical services, but the conditions in which they work. The causes of this malady are much more deep rooted than appear on the surface by a casual or superficial examination of the problem. Agricultural services have suffered from a number of chronic disabilities. The right type of environment for full blossoming of their initiative and constructive abilities have been lacking.

The obvious shortcomings and handicaps from which the agricultural services suffer are many.

(iv) Large Number of Temporary Posts

A very large number of temporary posts still exist in the States for long periods, despite the desire of governments that all temporary posts continuing for more than three years and having a reasonable chance of continuing should be made permanent. In some States posts sanctioned after 1932 have not yet been added to the cadres of the services, with the result that from 30 per cent to 80 per cent of the posts are still temporary in State departments. The pensionary rights of agricultural workers are adversely affected by the temporary nature of posts in the State Departments. The same is true in the Government of India also as far as technical posts are concerned. There are many cases where first rate workers have suffered severe hardships on this account.

(v) Low Scales of Pay do not Attract Talents

If the importance, which is due to the technical services, has to be given to them and the principle of parity between the technical services and the administrative services, as has been voiced by our leaders, is to be accepted, then it is essential that the technical services be paid at rates at par with administrative services. Unless this is done, the best of the talent available in the country, by and large, will be attracted to administrative services as at present, and starve the technical services of talent which is so essential for their healthy development. To quote the Joint Indo-American Team on Agricultural Research and Education : –

> "The disparity between salaries of scientists or teachers in agriculture and administrative positions is unrealistic and results in a down grading of scientific stature in India. Research and education are creative forces that provide new wealth for the country, both through the development of material goods and the enrichment of an enlightened population. The ablest minds trained in India's colleges and universities should be channellised into this highly productive technical research and education activity. This does not happen in the present circumstances, wherein the greatest opportunities for personal welfare lie in the acquisition of a position in the Government Administrative Services. A disproportionate share of India's intellectual competence is thus decanted off into the administrative field. While such services are important, it is unfortunate that the difference in salary and financial opportunity is such that many of those who might be more interested in following a scientific or technical professional career are lost to this field.
>
> It is important that the scales of pay of gazetted, as well as subordinate agricultural services should not be lower than those for other services, requiring comparable training periods, particularly, because the agricultural services do not enjoy the same social prestige, advantages and facilities that are available to other services. For deciding the scales of pay, the length of training required after matriculation should not be the only criteria, as subsequent prospects for promotion and prosperity in the service are also important consideration. In agriculture it is not uncommon for an officer not to secure a promotion till he has completed 20

> years service in the subordinate grade and then retire. In an important State, a majority of officers in Class I service in the scale of Rs. 350-1150, retire at a pay of Rs. 500 to Rs. 800 p.m. without ever reaching anywhere near the maximum of the time scale. A few years ago only recently the rules have been revised to a better scale."

(vi) Slow Promotions and Lack of Opportunities for Promotion

The chances of promotion to higher posts or higher scales of pay are also much less. Cases are not uncommon where even after 20-25 years of meritorious service, good workers are unable to secure promotion in the department. In some States promotion is based on a negative selection and as long as there is nothing against an officer in his personal record, he is given promotion according to his seniority. In another State, research and teaching posts are filled up only by direct recruitment and senior workers who exceed the age of 50 years are precluded from promotion. While all the I.A.A. Officers will get the senior scale of pay going up to Rs. 3000 and 33 per cent of the P.C.S. Officers are likely to get this scale, in the agriculture services only one out of 100 or even less, may get near about this scale and that too after putting in two to three times as much service as put in by the officer beginning to the general administrative services.

There are no selection grades in technical services, except perhaps in some States, and there too the grades are unattractive and the percentage entitled to the grade is extremely small. Delayed promotions among agricultural personnel, when their colleagues in administrative services or even in commercial organisations get rapid promotions are disheartening and lead to frustration. It is a paradox that services concerned with developing the agricultural economy of the country and food production, so essential for the success of our plans, should be paid the worst and have the least chances of promotion.

(vii) Promotion by Seniority alone would Not Do

The Indo-American team in their Report have observed:

> "Another major problem in the advancement of scientists is the transfer of specialists trained in a certain discipline to high level

positions in another scientific field for which they are not especially qualified, but to which they are considered entitled on the basis of seniority. An individual trained in palaeontology who took a position as a teacher of plant pathology, immediately following his advanced training, since this was the only post available at that time. After more than ten years in this post he obtained a position as a research specialist on sugarcane. Although he had no previous research experience, and no significant knowledge of sugarcane breeding and production practices, he was appointed to this position because of his seniority rights to the post. After several years in this position he was transferred to a post as an 'expert' in the breeding and improvement of cotton, wheat, barley and paddy. Eventually paddy was eliminated and he is currently an expert on wheat, barely and potato breeding. It was stated that this individual would have preferred continuing in the teaching profession but had no opportunity to advance financially to a reasonable level without transferring to another field to work. This is being slowly introduced in agricultural universities and Research Institute/Stations.

In a second instance, a scientist has been working for more than ten years on the breeding and improvement of oilseeds and pulses. His experience includes work with 26 different crops, all requiring specialised knowledge of their genetics and breeding behaviour in the improvement research programme. The head of this research section has recently taken another post. Instead of advancing the well qualified existing staff member, another man from outside of the section was being brought in because he was a senior individual. The net result will be that India will lose the competence and background of experience that has been developed through the many years of efforts of the scientist working in oilseeds and pulses since that man will eventually seek a position outside of his section to which his seniority will, in turn, entitle him."

Public interest or the interest of the subject-matter should be safeguarded by the retention of the good research workers and their promotion in the appropriate subject-matter. Men with competence and drive get things done in spite of normal difficulties that exist in any administration. Ability to show progress in spite of normal administrative difficulties is noticed in a few instances, while, in a majority of cases, this is lacking. If these virtues are to be promoted, there should be adequate recognition of merit at all stages, and those who have not shown initiative and progress should be made to realise that their advancement would not merely depend upon seniority.

(viii) Supernumerary Posts

Some system has to be devised whereby the staff which has the requisite aptitude and has acquired in a special subject, is continued in the same line without allowing their prospects of promotion to suffer. To achieve this, it is suggested that the creation of a new supernumerary post in each superior cadre. If an individual whose services are considered indispensable gets promotion in another line, it should be possible to retain him, on promotion, against one of the supernumerary posts, till such time as a suitable vacancy in the normal promotion and the supernumerary post could be released to the pool.

(ix) Opportunity for Leadership

The chances of leadership in agricultural services in our country are limited, if not non-existent. It is human nature to aspire to assume leadership and come to the top. This is as it should be, if we are to continue to progress. However, this is the destiny of only a few; but in a welfare and democratic state, the opportunities of becoming leaders should be equal for all. In our administrative set up, which now forms the base and scene of welfare administration in the country, the dice is loaded against the technical services. The mantle of leadership right from block, tehsil, district to the state level, in the present context of things, will invariably be donned by the members of administrative service. There are hardly any opportunities for technical services to work and aspire for this leadership, except when it may please providence to choose a solitary individual here and there. It will be a negation of social justice to confine leadership in the field of agricultural development to any one class or group of individuals.

The impact of poor prospects in agricultural services exerts an adverse influence in many things. Quite often, technical men find that officers from administrative services who have worked in a position subordinate to them, very soon find themselves in position of authority in the higher rungs of the administrative machinery, while they remain static on account of slow promotions and an absolute bar to their working in administrative posts in the field or the secretariat. This creates a psychological background which is not conducive to their contributing their best in the service of the country. Is it then any wonder that the technical services have developed certain complexes due to the environment in which they have been nurtured?

(x) Formation of an All India Agricultural Service

Nothing short of forming an All India Agricultural Service, with scales of pay and prospects at par with the Indian Administrative Service would raise the morale and efficiency of the agricultural services and thereby attract equally suitable material for manning the services and bring about uniformity of standard. The formation of such a service will also facilitate exchange of officers from one state to another and from the states to the Government of India. Such exchanges will not only help to produce officer of All India experience and competence, whose number has dwindled fast, but also unable the Government to attract and hold in the agricultural services some of the ablest minds trained in India's colleges and universities. An All India Agricultural Service in the past served the country well and helped to produce some of the top agricultural and animal husbandry specialists.

Financial stringency should not be made the basic consideration for deciding against the formation of an Indian Agricultural Service, considering that it did not stand in the way of the expansion of administrative services after the war, and more so after Independence in spite of the fact that they were numerically much superior than the Agricultural Service. Actually a bill for the Indian Agricultural Service was introduced in the Rajya Sabha and passed in 1971, but it lapsed as the Parliament was dissolved for fresh election. So that was the end of it. This should be revised.

(xi) Revision of Service Rules

Service rules have not been revised in some States for the past 25 years. In one State no recruitment rules exist for any post barring a few senior posts for which rules were framed in 1931. Nor is there any system of framing qualifications and method of recruitment whenever a post is created. Certain draft rules are reported to have been prepared and are under consideration, but even these rules do not seem to provide a quota reserved for departmental candidates and direct recruits. As posts are filled mostly by promotion, on the basis of seniority, there is no recognition of merit and there is no scope for the development of initiative, keenness or enterprise, because each man feels that his promotion will take place only by turn. The non-existence of recruitment rules and qualifications prescribed in advance, together with absence of published gradation list, can lead to *ad hoc* adjustment of recruitment procedures to suit a particular situation, which demoralises the service. It is understood that a number of posts have been made permanent recently and the problem of confirmation and promotion with the agricultural services. For this purpose, it is recommended that an experienced administrative officers from the Appointment Department be placed on deputation with the Secretary, Agriculture for drafting or revising rules of agricultural services, after taking into consideration the recommendations of the Agricultural Administration Committee.

a. Prescribing different qualifications for the same post on different occasions.
b. Alterations in the channels of promotions from one special subject to another e.g. from research to extension or *vice versa*.
c. Frequent changes in the seniority.
d. Ambiguity regarding the examinations to be passed, and the conditions to be satisfied before confirmation, promotion or crossing of the efficiency bar.
e. Inadequacy of the selection grade posts involving a higher start in the scale; for example persons appointed to the posts of Head of Sections, Deputy Directors etc. may start at a higher level as in the case of Supdt. of Police or a Supdt. Engineer.
f. Absence of supernumerary posts which can be given to outstanding research workers for enabling them to stick to their particular problem or line of research.
g. Absence of an adequate reserve in each cadre for meeting the leave, training and deputation requirements. On the basis of past experience it should be possible for the Departments of Agriculture to determine the average number of posts required for deputing officers to work in ICAR etc. The staff requirements based on Community Development pattern should also be added to the appropriate cadres.

h. Absence of procedures and qualifications for selecting departmental candidates for post-graduate and foreign training; panels selected by departmental committee should be available at all times.
i. Absence of procedures for periodical assessment of officers work.
j. Absence of procedures for the selection of non-gazetted candidates for promotion by Departmental Committee which inspire confidence.
k. Absence of proper balance between selection grade posts and posts in other superior and lower cadres.

(xii) **Aptitudes**

It is obvious that only those who have the aptitude for agriculture should be recruited in the Agriculture Department and since most of the Agricultural graduates are absorbed in the agriculture departments it is necessary to ensure that only those students who have real aptitude for agriculture are recruited to Agriculture Colleges. For Admissions in certain agriculture colleges, preference is given to those who come from farming families, but both the conditions are capable of so wide an application that they become inreality, ineffective. It is quite conceivable that a man who owns land may not also have an aptitude for agriculture. Evolution of suitable tests in aptitude and their adoption prior to admission to the Agriculture colleges is urgently called for.

(xiii) **Qualifications**

Posts in the agriculture department are largely filled by recruiting agricultural graduates, though in research and teaching branches some post-graduates having training either in chemistry, botany or zoology are recruited. It is desirable that research and teaching posts are filled up by the most competent persons as ultimately on the quality of work produced by these persons will depend the prosperity of the country to a large extent. For extension and Farm Management branches of the Department of Agriculture, an agricultural graduate is very necessary and a science graduate will be a poor substitute. In these two branches local experience would be considerable value and, therefore, for the post of agriculture extension officers at the block level, Agricultural Officers at the Tehsil, Sub-divisional and District levels. No higher degree should be prescribed than a degree in agriculture. In one state the appointment of Extension Officer at the Block level has been thrown open to science graduates also. Appointment of science graduates mean that the limited number of Extension Officers post available for agriculture graduates would be for a long time be occupied by persons who would not be competent to deal with field work in agriculture. Such misfits would not normally be replaced for the next 25-30 years. In future recruitments, steps should be taken to rectify such mistakes.

3

EVALUATION AND ASSESSMENT OF DEVELOPMENT WORK

(i) Unit for Evaluation Progress

The quality of agricultural technocrats should be continuously improved. There must be a small unit at the headquarters of the Director of Agriculture for evaluating and assessing the progress of new extension recommendations, land development schemes (reclamation, minor irrigation, drainage and soil conservation and schemes for the training and organisation of the farmers. This unit may carry out every year, detail studies – schemes which have been recently launched, and also undertake a broad review of additional 4-5 selected items of work. This will help the Department of Agriculture and the Government in revising procedures and avoiding pitfalls. This unit should serve as the eyes and the ears of the Department of Agriculture.

(ii) Supervision to Promote Follow-up

It is sometimes noticed that owing to pressure exerted for completing a certain quantum of developmental work and spending the amount provided for the purpose, adequate follow up is often neglected. A feeling generally develope, even in good officers, that their duty is duly discharged once they have completed their specified work or distributed the aids such as seeds, fertilizers, implements, insecticides and fungicides supplied to them. For proper use of these facilities by the farmers, it is necessary that they should be trained and convinced of the usefulness of the facilities and organised in such a manner as to promote their fullest use. For bringing about the necessary change in the outlook it is necessary to change the items of reports, so that the reports contain information about the number of works completed, as well as the area benefited. The emphasis should be placed more on the actual achievement and not so much on the potential created. The inspecting officers of the Agriculture Departments can during their inspections, assist considerably in emphasing these points by making the study of the follow up programme and the use to which actually a particular facility is placed. An intensive drive to promote follow-up work needs to be built up. In order to achieve this, it would be necessary to intensify supervision.

(iii) Need for Thorough Inspections

Inspection and supervision have been relegated to the background and have been replaced by *ad hoc* visits for solving immediate problems or for gathering urgent information. This seems to be due largely, to a rapid expansion in the agriculture department and a change in work loads arising from a change in emphasis to developmental world. Inspection notes of the agriculture officers deal mostly with accounts and audit irregularities and the procedure dealing with supplies of fertilizers and seeds, rather than the technological standards adopted to demonstration and extension and the special modifications required to suit the local conditions.

Apart from casual and *ad hoc* visits, systematic and thorough inspections covering all the facts of the work of the officers are necessary for the proper training of the officers inspection of the technical work at different levels, particularly in research and extension and development. Either the inspecting officers have too large an area to cover, or properly qualified staff is not available in adequate numbers for discharging this important work in the Agriculture Departments. Inspection notes do not normally give detail technical guidance based on the field observations of the field officers. The manner and system of technical inspections, it is understood, has been laid down in the departmental procedures, but their observance has not been insisted upon or checked at different levels periodically. These instructions need to be implemented. The work-load of the inspecting officers should also be studied. Complete and detail inspection in a smaller area will give results of much greater value than an inadequate supervision over a large area. Each technical officer should be responsible for a minimum number of visits to the units under his control. Such visits should be purposeful and pre-planned to provide solutions on the spot in time of inspection and should also be utilised to check up whether the unit as acted upon earlier advice and instructions given, and if so, on correct lines.

(iv) **Systems of Oral Reports**

The senior officers should be in continuous touch with the work in progress under their control by having a system of oral reports from immediately below them, on a prescribed date in a week or a fortnight. This will also provide an opportunity for officers at all levels to discuss with their senior officers, local problems, periodically, and help the senior officers to keep the work with which they are concerned under a continuous review.

4

TRAINING AND ASSESSMENT

(i) Orientation and On-the-Job Training

In the past, the officers of the Agriculture Department considered the training of their deputies and subordinates as an important duty. There was a realisation that they would themselves be judged from the work of their subordinates. This fast disappearing concept needs to be restored.

Where fresh agricultural graduates are recruited for filling the posts they have to be given orientation as well as 'on the job' training. The orientation training for a research worker or for a teacher should deal with the such aspects as the importance of particular field of study, the latest developments therein, and the critical analysis of the achievement in the field. In case of an extension worker even when he has training in extension, an orientation course is necessary for providing him with background information dealing with the needs and problems of the particular area and the progress made in solving them. Even old staff when transferred, for the first time, from one agro climatic region to another should benefit considerably from such courses. Such courses may vary in duration from 2-3 weeks and could be conducted, if so warranted, in the off season, at the Agricultural universities by the Extension Education Directorate where experts from all fields are available and also in the Directorate of Agriculture or in the office of the Regional Joint Directors.

It is essential that such training should be given by senior, experienced and competent officers as successful techniques and methods have to be communicated to the junior colleagues during the course. The job-training should be designed to provide a working knowledge of the different jobs which the trainee will be required to control or be familiar with. The training has to be necessarily, of practical nature as an under-study in the job concerned for a period which is considered adequate for the purpose. Such training may involve with indoor and outdoor works of different types. For on-the-spot training, the departments of agriculture should work out charts indicating the nature of the job, the kind of training and the duration.

A systematic programme of training should be prepared in such a manner that ordinarily before the officer is called upon to fill a higher post he receives an orientation and 'on-the-job' training to equip him fully for the new task. All 2nd line officers who are likely to fill up the 1st line posts should receive training. If arrears are allowed to accumulate in this respect it would become difficult to impart necessary training, as withdrawal of large number of officers at the same time is not considered either feasible or desirable. Training should, therefore, be a normal affair and the needs of the training should be reviewed once a year at a convenient time such as end of December. This review should relate to the officers who are likely to retire during the following 18 months, and the posts that are likely to fall vacant because of promotion and sanction, of new posts. It would be appropriate if a study of this kind is made for research, teaching and extension branches of the agricultural universities and research institutions.

The rapid strides that have been taken in agricultural research, make it necessary for the agricultural officers to keep abreast of the latest developments. With the changing sense of values, the old fashioned habits of regular and systematic personal studies are falling into disuse. So much research work is being published even in a special subject that it becomes difficult for a specialist to keep pace with the development in the field. Moreover, a specialist has to keep in touch with the subjects.

The Agricultural Departments should make permanent arrangements for training and keeping their officers up to date. These arrangements should be systematic and operate in an automatic manner. The contents of these refresher courses should be carefully planned and there should be opportunities for discussion at the end of each talk given by the research specialist or a senior administrative officer. The course may be of a week to fortnight's duration and organised at Agricultural Universities and Research Institutes.

(ii) Intensive Courses

In addition the post-graduate training facilities mentioned above it is necessary it is necessary to organise at the post-graduate colleges, short intensive courses of 4-6 weeks duration of which can be taken up during the summer vacations by the members of the teaching, research and extension staff. These courses should cater to the needs of individuals between 4 to 5 post-graduate colleges all the subjects should be covered. The information regarding these courses should be circulated at least six months in advance so that it enables the officers to obtain necessary leave of absence which should be granted freely. A note should be made of such training in the officers record, and due credit and weightage given to the examination passed after the training at the time of promotion. This is very essential for the teachers of the various faculties.

In the developing economy new programmes are launched from time to time, and to meet the situation, special training programmes have to be organised, such as special training for extension officers, agricultural teachers of secondary schools etc.

(iii) Regular Supply of Current Literature

While the staff employed at research and education institutions are well supplied with current journals, periodicals and other scientific literature, similar facilities for training the mind need to be strengthened at smaller farms and institutions and departmental office in order to keep the staff employed at these places fully abreast of modern developments and advancement in agriculture subjects. Journals, periodicals, abstracts, reviews etc. connected with the subjects should be supplied to them regularly. It is the duty of the Government to keep the staff fully educated so as to improve their knowledge of current affairs. The extent to which modern knowledge is imbibed and assimilated should be checked up through a system of examination held at the end of the refresher courses.

(iv) Need for Advance Planning of Training Programmes in the Foreign Countries

The need for training should be assessed on a long-term basis for each State and panels to suitable candidates for different subjects be kept ready. The States and the Central Governments should nominate for foreign training only those candidates whose names are on the approved panels prepared in advance. These panels should be kept handy with the Central Ministry of Agriculture so the full use can be made of the educational assistance offered by the foreign countries. With the long experience gained in this work, *ad hoc* selections and training programmes can be avoided. Officers after receiving training should be posted back to the subject of their specialisation so that the fullest advantage is obtained from the training given. It has been observed that in the past, sufficient attention has not been given to this aspect and even now.

(v) **Training Reserve**

In view of the growing pressure of work, the State Agricultural Departments find it somewhat difficult to spare the personnel required for the training. On the other hand, in the absence of training, the work often slows down and the pressure is further intensified. It is therefore recommended that in all the cadres of the Agriculture Departments where there are more than 8-10 posts in a particular subject, a training and deputation reserve should be created in addition to the normal leave reserve. In the past quality of material has been attracted to research and teaching courses and because of this it is necessary to concentrate first on the training of extension or the field staff of the Department of Agriculture.

(vi) **System of Examinations Necessary**

Periodical courses and examinations should be prescribed and the officers who fail to pass these examinations should not be confirmed, promoted or permitted to cross the efficiency bars. Emphasis should be placed as mentioned before on the passing of prescribed examinations and individuals who fail to pass should be made to loss. This is considered necessary especially in view of the prevailing tendency to attach little importance to these courses and to treat them are as holiday at the cost of Government.

(vii) **Technical Supervision**

Agriculture Departments, are often handicapped due to too great dependence on seniority for promotion arising from very limited direct recruitment of young and promising men of initiative to the gazetted services in the department. Special steps have to be taken to recognise and reward competence, initiative and keenness. It is often complained that though the cadres of agriculture services have expanded there is no proportionate increase in the out of turn. This criticism applies equally to research, teaching and extension branches. For recognising competence, initiative and keenness, it is necessary that there should be adequate and thorough technical supervision, at different levels, so the right officers are located, trained and their work appreciated properly.

(viii) **Assessment by Boards**

Except for the confidential character-rolls and recommendations made when the officers are confirmed or permitted to cross the efficiency bar, there is little evaluation of the work of the officers. Through inspection to judge the work of the officer have almost disappeared. Numerically, officers have increased so considerably that supervising officers are not in a position to study their juniors. Adverse confidential reports are often viewed as charge sheets to be contested and attempts are made to intimate and question the *bona fides* of supervision officers. In these circumstances, certain categories of officers of the Agricultural Department should have an opportunity of being assessed at least once in three years by a competent board which may include the concerned and other Heads of the sections. In the case of the gazetted officers, it might be useful for the State Government to arrange for an independent assessment by Boards of Agricultural Scientist and technical men of repute from different parts of India. As a result of this assessment, it may be necessary to provide a change of place for certain officers who might do better in different line. It should be seen that these reports deal with the quality and quantity of out of turn rather than the personality of the individual judged. These reports will also enable the Director of Agriculture to assess the work of his departmental officers.

(ix) **Co-ordination to be Dynamic and Purposeful**

Co-ordination of work whether in the multifarious committees or in the field, is a knotty problem. What is most urgently required is the coordination of mind and action. A tendency has been noticed that some of the co-ordinated bodies function in a routine manner and there is little evidence of a systematic follow-up action. Evidence is also lacking to show that these co-ordinating bodies do not meet often enough. Further more, co-ordination cannot operate successfully unless adequate authority is vested in the co-ordinating bodies at different levels. To ensure this the posts of Agricultural Production Commissioners have been created in the State. He not only co-ordinate the various activities of various departments concerned, so as to open up bottlenecks, but also ensures orderly execution of programmes. He is empowered to call a file from any department in order to watch its progress, call a meeting of the Secretary and Heads of Departments under him to find out why certain schemes are delayed and where, and inspect the progress in the field. He has the authority to write the confidential reports of the Secretaries of the Governments and Heads of Departments with which he is concerned. This system has various and obvious advantages.

The progress of plan scheme is generally checked through various progress reports obtain monthly, quarterly and annually from departmental officers. There is a general feeling that field officers have been burdened with far too many returns. In several States, as many as 30-40 returns have to be prepared and submitted to higher authorities every month. Co-ordination does not mean multiplicity of reports and returns. Wherever committees exist for purpose of co-ordination, they should meet frequently on predetermined dates, say 1st of every quarter or every month. Such meetings should receive priority over the official engagements. Representation on these committees should be personal and not through the subordinates. These Committees should concentrate on the complicated problems which are awaiting solutions, and to make them effective, sufficient power and authority should be vested in them.

There is a gross lack of co-ordination between irrigation and Agriculture Departments, both at the level of planning and execution in the field. This has already resulted in the irrigation resources not being fully utilised. The Department of Irrigation, Co-operation and Agriculture and the common extension agency of community development, should work as a team, with the sole object of assisting farmers to intensify production.

Each Superintending Engineer for Irrigation should have an agriculture officer, whose duty should be to serve as a liaison between the agriculturist and the Irrigation Department. In special irrigation projects, the State Department of Agriculture are often neither consulted or associated. In all irrigation projects the State Department of Agriculture should be taken into confidence from the very beginning and special agriculture officers should be attached for ensuring maximum utilization of water. This is a must.

5

BUDGETING, SANCTIONS, PROCEDURES AND DELEGATION OF POWERS

(i) Existing Procedures Retard Progress

The existing procedures of budgeting and the incurring of expenditure act as a deterrent to progress in the speedy execution of scheme in the agriculture sector. These procedures are generally applicable to the sectors of the Plan but in the field of agriculture specially, which has to deal with the production of food for the common man, their effect is particularly felt. This is because, in the sphere of agriculture unlike other sectors, the nature and the extent of operations are to a large extent dictated by climatic considerations, e.g., seasons, rainfall, etc. Particular seasons of the year are propitious only for a particular operations and certain activities cannot be taken up if the particular climate does not exist. Once commenced, the time available for such operations is also limited. It is, therefore, necessary before the onset of a season during which particular operations are to be undertaken, to mobilise all the resources needed so that they are commenced in due time and after commencement, there is no risk of their interruption. This will not be possible unless the appropriate financial sanctions are available in time.

There is considerable gap between the physical targets and actual achievements for various agricultural schemes. This shortfall has often been ascribed as being due to the late receipt of financial sanctions. The delay naturally limits the amount of work that can be done in a particular year. Seasons do not wait for sanctions and late sanctions, inevitably result impartial work being done or no work done at all. There are instance where sanctions for Plan Schemes have not been received throughout the financial year to which they related, although the proposals have been submitted by the heads of department at the beginning of the year. Such delayed sanctions also contribute to a state of affairs in which the bulk of expenditure budgeted for the year, amounting in some cases as much as 80 per cent of the provision, is incurred during the last quarter of the year. Though the funds in every case may ultimately be spent, the results are not commensurate with the expenditure because of its not being incurred at the appropriate time. "It must never be forgotten" observed the late Prime Minister Pt. Nehru, "that nothing is more expensive as delay". The concentration of expenditure during the last 3 months are mentioned above, is the anti-thesis of planning and is not conducive to the methodical and even implementation of the plan schemes. Means have to be devised to ensure that the expenditure is spaced out through out the year on as uniform a basis as far as possible.

(ii) **System of Budgeting**

Under the present system of budgeting for the Agriculture sector (which may be true or other sectors also), action to include the necessary provision for plan schemes eligible for Central assistance for a particular year is initiated some time in August — System of the preceding year. The authorities subordinate to the Director of Agriculture, at a suitable level, forward their recommendations to the Head of the Department, who after scrutinising them, transmit them to the Secretary to the Department of Agriculture. After approval by Secretary with such modifications as may be necessary, they are referred to the Finance Secretary, who returns them to the Agriculture Secretary after examination in his department. The schemes are then forwarded to the Government of India and are discussed in January-February between Planning Commission, Foods and Agriculture Ministry and representatives of the State Governments. The Schemes finally approved for implementation together with the provisions therefore and the share of the State Government and the Government of India in terms of loans and grants are then indicated.

After the Central Budget is voted by the Parliament, on a request by the State Governments sanctions are issued by Government of India after detail scrutiny of the schemes submitted in consultation with the Central Finance Ministry. The scrutiny at the Centre after the voting of the budget has been considered essential, as the schemes are not presented insufficient details at the time the budget provision is included. After obtaining the sanction of the Government of India, the State Department of Agriculture has again to apply to the State Finance Department for funds in respect of such schemes. A similar procedure is followed in respect of schemes financed entirely by the State except that these are not referred to the Government of India for sanction.

(iii) **Improvements Suggested**

The above procedure envisages consultation with the Finance Deptt. on two occasions, namely, before a provision is included in the budget and later after the voting of the budget and before expenditure sanction is accorded. While a strict scrutiny of schemes is necessary in the interest of wise spending cannot be dispensed with. They are of the view that the various stages of action from the formulation of a scheme to its implementation should be completed more expeditiously and unnecessary stages eliminated. It should be possible, for example, to reduce the number of inter-departmental references and expedite decisions if the system of personal discussions among officers was encouraged, in preference to the practice of referring matters from one department to another in writing. Conditions should be created in which a Plan scheme can be commenced immediately on the advent of the financial year, instead of being held to delays in the financial or administrative sanctions. In particular, in respect of schemes spread over a number of years, the issue of administrative and expenditure sanctions in the second and succeeding years should be regarded as formality only and not interfere with the continuing execution of such schemes while awaiting the sanctions. The issue of sanctions without undue delays can be achieved if the necessary technical and financial scrutiny of schemes is completed before the financial provisions for them is made in the budget of the year. This presupposes the preparation of the schemes, initially in greater detail than is done at present and it might be useful for the financial authorities who would scrutinise them to indicate before hand through instructions, the minimum data which should be invariably furnished before budget provisions is accepted. The technical and financial pre-budget scrutiny referred to above will naturally take some time and the suggestion, therefore, involves readjustment in the calendar that might have been laid down for the preparation and preparation of the budgets in the State Legislatures.

There may also be schemes in respect of which, either because of their magnitude, or because of other reasons, it may not be possible to complete a detail examination of the scheme prior to its inclusion in the State budget. In such cases, a lump sum provision may be made and expenditure sanctions issued after detail scrutiny in consultation with the State Finance Department. Pending such sanctions the State Government can authorise the incurring of expenditure on preliminaries subject to certain financial limits.

If the above recommendations are accepted and provision in the State Budget is ordinarily made only after detailed scrutiny of the Plan schemes, there should be no necessity for a second reference to the Finance Department for expenditure sanction. It will be possible for the States to accord administration approval immediately after confirmation is received from the Government of India of the acceptance of a particular scheme for the Central Assistance and the Plan provision, so that all preliminary work is undertaken in respect of schemes financed entirely by the States, such approval may be accepted on the basis of their inclusion in the budget. This will enable the actual implementation to be commenced on the 1st April, the expenditure sanction being accorded, as soon as the budget is voted by the State legislature.

(iv) Revised Procedures

In respect of State schemes involving Central assistance one of the factors which had contributed to the delay in their implementation in the past had been the time taken in obtaining the Government of India's sanction to such schemes and particularly to the Centre's share of expenditure. These sanctions were accorded after detailed scrutiny of the scheme from the technical and financial angles. Under the revised procedure which has been introduced and will apply *inter alia* to the agriculture sector, the schemes in the State Plans for which Central assistance is to be made available will be sanctioned by the State Government under their normal procedure and no financial sanction will be issued from the Centre. A substantial portion viz. 3/4 of the assistance to which a State Government is entitled is made available as ways and means advanced by the Central Finance Ministry in nine monthly equal instalments commencing from May. These advances will be cleared later in the year by issue of formal sanctions to loans and grants.

(v) Organisational Improvements

As a further step towards improving the system of budgeting and expenditure control the Committee recommend that certain organisational improvements may also be undertaken. At the Centre, each administrative ministry now has an Internal Finance Adviser to render financial advice generally and in particular, assist in the formulation of proposals in the budget from their inception. A similar system was adopted by the States and Financial Advisers were posted to the State Departments of Agriculture as part of establishment, the financial processes connected with the formulation and implementation of Plan schemes would be considerably expedited. It is, however, essential that this adviser should have definite power to take decisions on behalf of the Finance Department. His appointment should facilitate expenditure disposal and he should not be permitted to become an additional channel of approach to the Finance Department. Under the present system of financial scrutiny, Plan Schemes are scrutinised in State Finance Department by officers who have had to actual experience of development work. Without questioning their ability, it would be of great advantage if the officers of the Finance Department charged with the scrutiny of such schemes have actual experience of the formulation and implementation of the Plan schemes in one capacity or another.

(vi) Implementation suffers due to Lack of Powers

The implementation of projects has suffered due to the lack of powers of the officers entrusted with such implementation. The tendency appears to be to concentrate all powers in the hands of a few officers on the top. Such concentration naturally, impedes the execution of schemes, as considerable time is lost even in the case of duly approved schemes in obtaining sanction by individual officers at the different levels, because their normal financial powers in regulations and instructions are limited and they have to approach higher authorities for relation of rules every time expenditure exceeding the limit laid down in their financial powers has to be incurred. There seems to be also some confusion as to whether after a scheme has been sanctioned, it is still necessary to obtain sanction again for the individual component parts of the scheme.

(vii) **Delegation of Powers at All Levels**

In the interest of speedy execution of work, there should be the greatest delegation of powers at all levels from the Director of Agriculture downwards. The powers that could be exercised only by the senior officers should be clearly specified and these might be added to or amended on the basis of experience. In respect of matters not specifically reserved for consideration and decision at higher levels, the officer implementing a programme, irrespective of position in the hierarchy, should be enable to operate on his own without making frequent references to higher officers. The existing rules and regulations governing the financial powers of different officers were framed several years ago with reference to prices prevailing at that time. With the increasing cost of living and the rise, 6 to 10 times, of the cost in the commodities, it is essential that these powers are revised upwards. Some of the States have already undertaken a review of the relevant financial regulations. This practice, however, is not uniform. If development work has to be executed expeditiously with the expenditure spread out evenly during the year, it is essential that the financial powers of officers at all levels are examined urgently and, where necessary revised so as to conform to existing prices of commodities and changing needs of the expanded development programme under taken by the country. This would be a matter to be taken up by each State urgently. In under taking this revision, the following general observations may, however, be borne in mind –

(i) Liberal powers for the purchase, sale, write off, and repair of tools, equipment materials (seeds, fertilisers, plant protection materials, apparatus and chemicals) should be granted to the officers of the Agriculture Department. Certain restrictions which are operative in other departments may have to be relaxed in the case of agricultural departments.

(ii) Procedures for purchase and sale may be prescribed and so long as these procedures are followed it need not be necessary to obtain a sanction from higher authorities subject, in the case of purchases, to the funds having been provided in the budget and the items being of very common use viz., seeds, fertilizers, manures, cattle feed, oil cakes, implements, fungicides, and insecticides.

(iii) Administrative powers to grant casual and earned leave, and to inflict minor punishment including censure and warning should be generally vested in the immediate superior and appeals should not be beyond the level of the second superior officer.

(iv) Authority to countersign bills and travelling allowance bills should vest in the first gazetted superior.

(v) All heads of sections have full powers to transfer their non-gazetted subordinates within their normal jurisdiction.

(vi) Existing procedures regarding the sanction of purchase of bicycles, typewriters, calculating machines, office clocks and furniture needs simplification and liberalisation.

(vii) Powers to rent godowns should be liberal for the officers who are required to purchase, stock and large quantities of materials. These officers should have powers to incur handling and transport charges according to prescribed methods or scale.

(viii) The officers in charge of farms, laboratories and workshops should have authority to carry out minor urgent repair in Government buildings within certain ceilings and funds should be placed regularly at their disposal.

The powers delegated should be such that the officers main task is not impeded because of lack of powers.

After the above delegations are made, steps should be taken to codify all financial and administrative powers not distributed over different books and executive instructions in brief, officers, supervisory, executive or otherwise, at the levels should have a clear indication of their financial and administrative powers.

(viii) **Restrictive Orders to be Revised**

Even in regard to the powers that have been delegated, the officers concerned felt circumscribed in the exercise of these powers as it was hedged in by so many restrictive issued from time to time that for all practical purposes they nullified powers given to them by the standing regulations such as the General Financial Rules. Even after the Financial powers are revised upwards, the issue of executive instructions from time to time will act as a strangle hold on the exercise of the normal powers. It is not surprising that there is a growing feeling among even responsible officers that Government was not trusting their officers enough. This feeling was not conducive to officers taking initiative and responsibility and should be removed; the delegation of additional powers should be both in letters and in spirit.

(ix) **Ad hoc Financial Powers**

In addition to the enhancement of normal financial powers of officers, each officer can be given, as an *ad hoc* measure, power to execute specified scheme for which budget provisions is accepted, without obtaining further sanction on any item of expenditure under that scheme even though such expenditure may exceed his normal financial powers. Such *ad hoc* authorisation may be effective during the particular financial year and given absolute power to certain officers on important schemes within the ceiling of the budget provision and subject to their observing other codal formalities. This procedure could be particularly adopted to speed up plan schemes which are lagging behind for various reasons.

(x) **Model Agricultural Organisation to Suit Present Needs**

If the agricultural departments are to fulfil the expectations of the people in the field of agriculture it is necessary that they must be organised to handle promptly and in an efficient manner the growing demands of the agriculturists. The departments should not only provide facilities for educating and training the farmers, but also for training the personnel required for manning the services connected with agriculture. Agricultural departments should demonstrate to the farmers better methods of cultivation and disseminate knowledge about these methods through various audio-visual channels and organise the farmers for adopting the improved techniques of crop production and working together for purpose of common benefit in activities such as land development protection of crops, marketing and organisation of self help programmes. The department should have the capacity to pick up the problems of the farmers and pass them up quickly to the research section for the prompt solution and rapidly transmit their solutions back, not only to concerned farmers, but to other farmers as well. The department of agriculture in the past have regularly secured and supplied certain quantities of seeds, fertilisers, implements, agriculture machinery and plant protection equipment; but this in not the legitimate role of the Agriculture department though this is certainly the function of the welfare state. It is necessary for the Agriculture Extension staff to concentrate solely on educating and organising the farmers to produce more. The department should deal with technical knowledge and education rather than trade in supplies. The field for extension is as vast as for agriculture research and the Department of Agriculture should equip themselves for meeting the demands for improved agriculture techniques which are developing fast in all areas.

(xi) **Organisation of Agriculture Information**

The success of agriculture programmes depends upon among other things, on a suitable farm Education Service reaching the cultivator and extending to him the current knowledge of improved agriculture practices. A Farm Education Service which is intended to educate the farmer in adopting the improved techniques of farming has three essential components, namely (i) Farm Advisory Service, (ii) National Extension Service and (iii) Agriculture Information Service. The Government of India and the State Government are developing effectively National Extension Service at the Block and Village levels which will be expended eventually to cover the entire country. The village level worker is intended primarily to work at the village level, but he does not possess adequate training in agricultural practices on scientific lines, although a major portion of the farmers occupation relates to agriculture. He probably knows less about "agriculture" than many "good" cultivators. His effectiveness in improving agriculture and his utility to influence farmers will increase if he has constant supply of good education material and current scientific information, produced in the manner which could be understood by him, for transmission to the cultivator. A suitable Farm Advisory Service should be developed in each Agriculture Department so that properly trained Extension Specialists in specified subject-matter fields, who will be free from the regulatory and enforcement functions which now form the major portion of the duties of the officers of the State Department of Agriculture, would form an effective link between the research institutions and the field level extension Agency. This has been developed to a great extent, but requires strengthening.

The Farm Advisory Service to be fully effective should be supported by an Agriculture Extension Service, consisting of trained staff of Information Officers specialising in mass communication methods. Mass Communication is just as specialised field as Agriculture Research or Agriculture Extension. Neither a research worker nor an extension specialists could be expected to prepare extension publications, leaflets, pamphlets, charts, exhibits, visual aids which form an important plank in conveying agriculture information to the farmers. While the research workers and the extension subject-matter specialists of the Farm Advisory Service would provide scientific information to processing of such scientific material into popular extension publication, visual aids and other information data, will be the special responsibility of the officers of the Agriculture Information Service. Steps were initiated during the 2nd Five Year Plan to develop Agriculture Information Units for this purpose in the State Departments of Agriculture. Similar Information Unit has been developed in the Extension Wing of the Ministry of Agriculture, with which the State Units will work in close cooperation, but the State Units will have to pay special attention, to the production and distribution of information literature in local languages and in filing in gaps in scientific information having a local bias. It is important that agriculture information work should be distinguished from publicity and propaganda work which is the normal function or the Information and publicity departments of a State. Full-fledged Agriculture Information Bureaus should be developed in the Agriculture Departments principally to deal with scientific information to reach the farmers and farm women with knowledge of 'how to increase his/her income'. In any State where agriculture information work is mixed with publicity and propaganda, it should be separated and located in the agriculture department to be developed into an agriculture information bureau to deal with extension type leaflets, pamphlets, journals, visual aids, films, exhibitions and all means of mass contact with the villager. The work is a specialised one and the presentation of the material in a language and form acceptable to the Indian Villager should require more and more recognition. The material produced in this bureau should be extensively supplied to the field extension workers such as the Village Level Workers and also to the farming public. The bureau should be under the charge of a Joint Director of Agriculture (Agriculture Information). He should be adequately provided with staff, equipment and funds to effectively carry out his work. This is lacking in many States, as such should be strengthened. There should be also very close link with the Extension Education Directorate of the State Agricultural Universities who have also some expertise and funds. If the Agricultural Universities require assistance then the State Government should assist them.

6

AUTONOMY OF ICAR

A quick and concerted approach should be made to streamline the existing operational procedures for according real autonomy to ICAR and immediate steps should be taken by the Council to develop its rules and regulations in a manner that will ensure its complete financial and administrative autonomy.

(i) ICAR Society

ICAR Society is the Supreme Body and is responsible for providing policy directions to the governing body and other Constituent Units of the society, as also to review the progress and performance of different constituent units. The Group considers that in order to ensure effective functioning of this Supreme Body, it should be provided with a restructured Governing Body which could assume overall responsibility for providing superintendence, direction and control overall affairs of the Council. The society could meet twice a year and should have well structured agenda with detailed notes which should be available to the members at least four weeks before the date for the meeting.

(ii) Governing Body and ICAR Headquarters

The Governing Body should be a small and compact body composed of eminent professionals, holding expertise in agricultural research and education and capable of leading the council to achieve higher and yet higher goals. The Governing Body should, in turn, be assisted by a strategic Planning Committee to advice on major policy issues related to Agriculture Research and Education, finalisation of new research thrusts and their *inter se* prioritisation and allocation of resources. The Scientific Panels may be given a broader function to review research programmes in a major discipline/area and assist the

Policy Planning Committee. The existing Regional Committee may be converted into Agro-Ecological Zones Advisory Boards in order to monitor progress in research and training in the 15 agro-ecological zones determined by the Planning Commission. These Boards should also identify critical gaps in on-going efforts in agricultural research and education in the region.

(iii) **Subject-matter Divisions**

The Governing Body should execute its policies through the Director-General, who is the Chief Executive of the Council, assisted by other Subject-matter Divisions and a Secretary of the Society in charge of Administrative and Services Division (comprising Personnel, Finance and General Administration). In view of the growing importance of international co-operation in agricultural research and education which is (i) there should be a strong technical division in International Co-operation headed by one of the DDGs. The Subject-matter divisions in the Council's headquarters should be under the charge of Senior Managerial Scientists of the status of Deputy Director-General, who should be assigned exclusive responsibility for national level planning and finalisation of the major research and education programmes of their division and their allocation to different Institutes according to their mandate and to the SAUs as part of AICRPs/network projects keeping in view the needs and strength of existing infrastructure, (ii) to review and monitor the progress of the implementation of these programmes, (iii) to allocate financial resources to the different programmes and to ensure their optimum utilisation and exercising complete freedom for re-allocation of funds from one primary unit to another within the Institute and from one Institute to another institute under their charge, (iv) to provide required superintendence, direction and control over the activities of the Institutes, particularly with regard to relationship with the Headquarters and other outside agencies, (v) to provide liaison between the research institutes of the Council and the SAUs with regard to co-ordination and implementation of research programmes and also to identify demarcation of research programmes and also to identify demarcation of research programmes and also to identify demarcation of R & D responsibilities, (vi) to act as Divisional Heads at the Headquarters and to be fully responsible to the Director-General for the activities of the Institutes under their charge, (vii) to exercise financial and administrative powers to the same extent as vested in the Director-General, of course under the overall policy framework decided by the Director-General from time to time.

(iv) **Administrative and Services Division**

As suggested above, while the subject-matter divisions should be under the charge of Deputy Director-General, the Administrative and Services Division should be under the charge of the Secretary. At present there is no post of Secretary, ICAR in the Council but the functions of this post are discharged by a Joint Secretary to the Government of India in the Department of Agricultural Research and Education with *ex officio* status as Secretary, ICAR. It would be more appropriate to provide for the position of Secretary is solely responsible to the Society. The functions of the Secretary should be clearly redefined so as to enable him/her to discharge the functions as listed for the Secretary of any Society. This is particularly necessary because the Deputy Directors General should be the Heads of the subject-matter division, both in respect of administrative and scientific spheres. This will necessitate redefining the functions of Secretary, ICAR so as to assign him/her total responsibility in respect of (i) Headquarters establishment, (ii) to deal with all policy matters relating to administrative support to the scientific managers in the divisions, (iii) to act as Principal Adviser to the Director-General in all administrative and personnel matters, (iv) to discharge all other functions of the Secretary to the Society as laid down in the rules and regulations and bye-laws of the Society. Such a step is all the more necessary because of suggesting the complete delegation of powers to the Institutes and their Management Committees which would apparently reduce work of the Headquarters to the absolute minimum and will be confined only to policy and budgetary matters.

(v) **Research Institutes**

Each ICAR Institute should operate with a specific and well-defined mandate and an identified set of objectives. The mandate should be defined keeping in view the changed agricultural research scenario in the country with Regional Research primarily being undertaken by the SAUs. Further the redefined mandate should be within the framework of the overall research planning and policy of the council, as determined by the strategic planning committee. Each Institute will have a Research Committee with Chairman and Members as eminent scientists from outside the ICAR system in major disciplines of the Institute and Director as Member Secretary to scrutinise specific projects for approval and financial support and their continuing evaluation and monitoring.

Directors of the Institutes, directly or with the approval of Management Board/Committee should enjoy full powers, both financial and administrative, within the budgetary grant and specific items of responsibility. The institutes will also have the responsibility of recruitment and assessment of all positions except Research Management positions. The ASRB will be responsible for conducting qualifying examination for suitability to entry to Agricultural Research and Education positions in the ICAR and State Agricultural Universities (SAUs). The financial powers will be delegated down the line up to the Project Leaders.

(vi) **Norms and Accredition Committee**

This Committee which is charged with the responsibility of looking into academic matters, should be reconstituted and provided with suitable authority to ensure effective role of ICAR in determining and maintaining the standards of agricultural education. ICAR should also strive to attain suitable authority for this Committee in order to make its functioning more effective. This Committee should follow procedures similar to that of UGC with respect to funding Agricultural Universities.

(vii) **Management Committees/Boards**

Each Institute of the Council has a Management Committee/Board, to broadbase decision-making on all Institute matters. These bodies are currently more of advisory nature rather than as Executive Bodies vested with full administrative and financial powers. Keeping in view the basic mandate of granting autonomy to the research institutes of the Council, it is necessary that full technical, financial and executive powers should be vested with the Institute Management Committees. Once the budget is allotted for specific items to the Institutes by Headquarters after careful scrutiny and approval of the

Governing Body, complete authority should be given to these Committees for their functioning without any reference to Headquarters as also to decide matters falling outside the purview of the delegated powers of the Directors. These Committees should in fact function like Management Boards of the Agricultural Universities.

(viii) **Headquarters Functions**

The Director-General, who is the Chief Executive of the ICAR, manages all the affairs of the ICAR with the assistance of the Technical Divisions and the Administrative Wing at the Headquarters. On the technical side, the Director-General is assisted by the DDGs, ADGs and senior scientists; on the administrative side, he is assisted by the Secretary, ICAR, and other secretarial staff. The Institutes are reviewed, guided and serviced by the Headquarters on all matters. During our discussions with a number of scientists and others, we notices a general feeling that the present dichotomous structure at the Headquarters has resulted in delays and sometimes even inconsistency in directions, thus affecting adversely the smooth-working of the institutes. We have carefully looked into this aspect and feel that such a structure does not exist in any of the similarly placed research organizations. Further, any technical proposal does involve an element of administrative and financial consideration and cannot be dealt with in isolation. Such a clear distinction between the administrative and technical wings is bound to give rise to two distinct lines of command at the ICAR Headquarters, as far as the functioning of the Institution is concerned. We, therefore, recommend that this dichotomous structure should be dispensed with. We also recommend that there should be a complete integration of the technical and administrative wings at the Headquarters.

It has also been noticed that the Headquarters is currently dealing with trivial matters concerning administration and accounts, many of which are of a routine nature. The Institute should be provided with sufficient autonomy of functioning so that only very few cases need be handled at the Headquarters level.

The ICAR should adopt modernization in its office procedures and equipments that will cut down delays and improve efficiency. For this, the ICAR should immediately introduce officer-oriented system of working at all levels with adequate provision for level-jumping, as also Desk Officer system in place of the existing conventional sections.

The functions of the Secretary, ICAR, may be revised to enable him to devote more attention to the organisational matters and extending advise to other scientific and administrative functionaries. The primary responsibility of the Secretary, ICAR should be –

(a) To act as an Adviser on all administrative matters to the DG and to extend his advice to the DDGs/Directors where sought for,
(b) To act as Secretary to the Society, the Standing Finance Committee and the Governing Body and to organize their meetings and process the implementation of their recommendation,
(c) To oversee administrative, vigilance and organisational disciplines as also to deal with representations and appeals falling within his own powers and the powers of the DG,
(d) To deal with establishment matters in respect of the Headquarters' staff,
(e) To finalise budget estimates and supplementary estimates of the ICAR as also to deal with Audit Reports and Appropriation Accounts,
(f) To enter into agreements on behalf of the ICAR, and to represent or cause to be represented on behalf of the ICAR to disputes before the Courts/tribunals/arbitrators.

7

VIEWS OF THE PRIME MINISTERS, SENIOR OFFICERS AND SCIENTISTS OF INDIAN COUNCIL OF AGRICULTURAL RESEARCH

(i) Views of Prime Ministers, Chief Ministers, etc.

The late first Prime Minister Pandit Jawaharlal Nehru with whom I discussed this subject of technocrats/Scientists being made the secretaries of the technical ministries, he agreed on principle and even said that a few have been appointed like Dr. S.S. Bhatnagar. He wanted to do more, but generalist put lot of obstructions since they were running the machinery of the Government of India so the beginning was made, but progress became very slow.

I discussed the same issue with former Prime Minister late Smt. Indira Gandhi. She was keen on inducting technocrats/scientists as secretaries in the various technical ministries and departments, and she did it. In her time many of them were appointed, but yet the progress was slow.

Similarly I discussed the matter with former Prime Minister late Shri Rajiv Gandhi as late as in October 1988 when Golden Jubilee of *Assam Tribune* was celebrated in the Vigyan Bhawan, New Delhi. He said at that time Government of India have appointed over 14 technocrats/scientists as secretaries and will appoint more. But, I reminded him then that out of our 44 ministries/departments this small number of 14 or so is too small. Why a technocrat cannot be appointed as secretary in Agriculture Ministry as it was there for about a year when Dr. M.S. Swaminathan was appointed. He said he will consider but destiny has taken him away. Let us see how far we can go in our effort.

Plate 4. Pandit Jawaharlal Nehru First Prime Minister of India with General Srinaghes, the Governor of Undivided Assam along with author at the Raj Bhawan, Shillong.

Plate 5. Srimati Indira Gandhi, Third Prime Minister of India and the author, explaining the Policy of India to the young farmers being invited to the U.S.A.

Plate 6. Sri Rajiv Gandhi answering the author's question on Appointment of Technocrats as Secretaries to the Technical Ministries of the Government of India.

Similarly I asked Mr. P.K. Mohanta, former Chief Minister of Assam. He said this is a good idea and he will examine. It remained there.

They all realise the importance of making technocrats/scientists as secretaries, additional secretaries, joint secretaries in the technical ministries/departments. It is for us to pursue it aggressively till we can assert our rights. This requires co-operation of all and hard work for technocrats/scientists.

(ii) Views of the Senior Officers and Scientists of ICAR and Ministry of Agriculture

This is what Dr. N.S. Randhawa, former Director-General of Indian Council of Agriculture Research said. He got some co-operation from the bureaucrats working in the ICAR under him, but not full co-operation. When he was asked whether is it a fact that the Financial Adviser to the ICAR created difficulties. He, then replied in the affirmative. When I told him that some of the Deputy Director Generals mentioned to me that the Secretary, ICAR, the deputy secretary and under-secretaries turned down their proposals on technical projects on flimsy grounds and they were not at all happy with the working of the ICAR on a dichotomy basis. Then he said that such things have happened, which should not have been done. On the whole my impression was that he suffered from the interference of the bureaucrats in the ICAR and instead of helping and pushing through the programmes they created more problems only. As such he agreed with the recommendations of the ICAR, Revised Committee of 1988 and also agreed with the recommendation of the Report of the working group on Agricultural Research and Education for the formalities of Eighth Five Year Plan of August 1989.

(iii) Views of Dr. R.M. Acharya, former Deputy Director General, ICAR, Animal Science

a. The Administrative matters should be managed by professional managers who are scientists/technocrats trained in management.
b. The Financial Administration should be managed by Scientists Technocrafts who are trained in Financial Management.
c. He did not get any help from the bureaucrats including their staff both in administration and Financial matters. Member

Finance of ICAR (Additional Secretary Finance) and Financial Adviser of ICAR did not help him.

d. ICAR has the dichotomy in administration which is standing on the way of progress and implementation of the programmes of ICAR.

He fully agreed with the recommendation of the Report of the ICAR Review Committee of 1988, and also with the recommendations of the report of the Working Group on Agricultural Research and Education for the Formulation of Eighth Five Year Plan of August 1989.

(iv) **Dr. C. Prasad, Deputy Director General ICAR (Extension)** stated that:

1. The Scientists/Technocrats should be trained to be good managers by imparting regular training on administration, personnel management, and Financial management.
2. A bureaucrat cannot be trained in Science/technology whereas a technocrat/scientist can be trained in personnel management and financial management. ICAR has a large number of bureaucrats with supporting staff.
3. Bureaucrats have been very indifferent to the proposals submitted by the scientist/technocrats and always want to demoralise them.
4. The bureaucrats of the associated finance withheld many projects by making unnecessary queries and sometimes the validity of the important projects.
5. In ICAR the bureaucrats have to be approached for various services of the Secretariat and in this process 50 per cent time is wasted and delay takes place. It took one and half year for getting some additional aid, but it had to be shelved because of the delay. It is a sad state of affairs. These situations can easily be corrected.
6. The dichotomy in ICAR is rampant and for the growth of Agricultural Research and Development it should be changed. An instance at point was a transfer of an officers took lot of time of D.G., DDG. (Ex) and Secretary ICAR. It happened because of dichotomy. This process delayed the progress of the programme.

Dr. C. Prasad also agreed with the recommendations of the ICAR Review Committee of 1988 and recommendations of the working group on Agricultural Research and Education for the formulation of Eighth Five Year Plan or strategy for implementation of the Eighth Plan of August 1989.

I have met some more Deputy Director Generals, many Assistant Director Generals, Director of the ICAR institutions, Senior Scientists. All these who have worked in the ICAR system for more than 15 to 20 years agreed with the recommendations of the ICAR Review Committee of 1988 and the other Committee of 1989.

ICAR Review Committee was a very high power Committee appointed by the President of the ICAR Society who is also the Union Minister of Agriculture with Dr. G.V.K. Rao, former Secretary Agriculture in the Ministry of Agriculture, Government of India, and Member Planning Commission as the Chairman in 1987, and they submitted the Report in 1988.

(v) Recommendations of the Working Group on Agriculture and Education for the Formulation of the 8th Five Year Plan Implementation of Personnel Policies

The Committee has emphasised three dimensions of the personnel policies, namely (1) the effectiveness of the ICAR depends upon the quality and quantity of work done by its scientists. The system should adequately support the scientists through its technical, administrative and supporting services. (2) The manpower needed by the organisation should be acquired and sustained through a system of recruitment of the best available talent in the country. The policies should provide for professional growth, provide reward to the scientists on the basis of their professional achievements, and manage the cadre of manpower in such a way that the total scientific manpower is able to meet the challenges of the emerging needs of agriculture which are getting highly diversified and changing rapidly. (3) The Personnel policies should constituted a vital aspect of management in the ICAR system beginning with manpower planning, recruitment, continuous training and professional growth of the scientists. All this has to be operated as a chain of related activities, each one of it affecting the other and being affected by them. At present this seems to be disjoined instead of being integrated so that at every step the feed back and the required adjustments may not be possible. In view of the fact that this is an important area of professional management in sciences, the Committee was the opinion that the DDG (Education) should be incharge of this activity. While the ASRB which belongs to the ICAR system will work independently for the purposes for which it has been set up. The DDG (Education) in addition to the present responsibilities, will function to provide feed back to the ASRB and articulate the work of the ASRB into the overall personnel management of the ICAR System. The present unit of the Director (Personnel) at the Headquarters will work under the DDG (Education). The ICAR at its headquarters has till now been developing only in scientific areas, such as Crop Sciences, Animal Sciences, etc., which represent only a vertical line in the management of the organisation. Too many such vertical lines dealing with subject areas are not conducive to the integrated management of the research organisation. What is required now is a horizontal line of management activity getting across the disciplines. The management of personnel policies represents such a professional activity which should be headed by a scientists of repute. This will provide to the Research and Analysis Wing in the ASRB a large amount of data which will enable a better personnel management and even help develop newer procedures. The Committee emphasised the urgently of this need so that an important sector of management at the Headquarters level is appropriately attended to.

The working group on Agricultural Research and Education for the Formulation of Eighth Five Year Plan was constituted by the Planning Commission on September 30th, 1988 under the Chairmanship of the distinguished Scientists, Dr. M.S. Swaminathan, former Director General of ICAR and also of International Rice Research Institute, Phillippines.

This distinguished working group made very positive, practical and sound recommendations in 1989 which it is understood the ICAR has accepted them and working out the details of their Eighth Five Year Plan based on these recommendations.

(vi) **Dr. A.C. Chatterjee, Former Animal Husbandry Commissioner**, who retired only in April 1990 said that bureaucrats did not allow to function properly, and those who were his superiors knew very little of Animal Husbandry but yet pretended to know all. Thus created confusion.

According to him even when policy decisions were taken the technocrats were not taken into confidence, and did not discuss with them. When proposals were prepared for the approval of the Union Minister and Finance he was obliged to send to Additional Secretary who was a bureaucrat, and when the Animal Husbandry Commissioner was himself in the rank of the Additional Secretary. Why?

The Animal Husbandry Commissioner was not authorised to send his junior officers on tour, and he had to forward those proposals to the Additional Secretary, which he and other officers thought as an absurd situation. I also think that was readily making the A.H.C. practically non-functional.

Dr. Chatterji said that the implementation of the plan projects were hampered very much. The bureaucrats did not allow them to function properly.

Since April 1990 Division of Animal Husbandry was headless, as the Ministry did not select a successor ahead of time, neither the ministry allowed him to hand over the charge to the senior most Joint Commissioner. But the Division was put in-charge of a Joint Secretary who is a bureaucrat without any knowledge of animal husbandry. This is a very unfortunate, and for such an important division. The Ministry should always think ahead of time, select a technocrat to take over immediately. The new Animal Husbandry Commissioner (A.H.C.) joined only in June 1991, after over one year.

Dr. Chatterji stated that 12 senior posts are lying vacant as on 15th June 1991 due to faulty recruitment rules. This was also the view of the other officers of the division. The A.H.C. should have been given full control of the Division and he also should have exercised his rights and authorities. It seems A.H.C. also did not pursue vigorously.

It was suggested by the A.H.C. and other officers that all senior and junior officers should be sent for regular training in their own subjects, and also for training in personnel management, extension management, training management and financial management. This will increase the competence of the officers.

According to Dr. Chatterji, A.H.C. that it was only in the time of Mr. C. Subramaniam when he was the Union Minister that technocrats were given their due weightage, and made them responsible of the divisions concerned.

Senior technocrats of the Ministry stated that the technocrats are not taken into confidence while taking policy decisions. The bureaucrats take the decisions. They agreed with the views expressed by the Animal Husbandry Commissioner.

In the case of Horticultural division of the Ministry, the Division has not been strengthened with technical persons since 1981, after it was separated from the crop division. The recommendation of the planning commission on Horticulture was also shelved by the bureaucrats of the finance division. As such there is frustration amongst the technical staff. The post of Horticultural Commissioner has not been upgraded to the level of A.H.C. and A.C. The post is also lying vacant since April 1991.

The Agriculture Ministry including all its allied departments should be given highest priority in all matters. So also in the States. This includes all the technical departments should be in priority list.

(vii) Views of Senior Technocrats of the Ministry of Agriculture

A number of Senior technocrats of the Ministry of Agriculture were interviewed to find out their views and to get their suggestions. They stated that the views of the technocrats are not taken in policy matters. They are used as a note writer like general assistants, very little power, both administrative and financial were also delegated to them thus making them ineffective. This is leading to frustration and poor implementation of the programme. It has been stated that bureaucrats very often interfere in the implementation of the programmes. The bureaucrats who are posted as heads of divisions take atleast a year to learn the work of the division, and by the time they get experience and learn the work they are transferred, as such there is no continuity.

In order to make the technocrats more competent there should be a regular policy and programme for training in their own subjects as well as in administration, personnel management, and financial management. There should also be regular programme for referesher training as it is for the bureaucrats, which they must attend. Similarly such refresher courses for the technocrats should be organised. This was expressed by all the technocrats. They also expressed that they should be sent for refresher courses in their own subject regularly.

The salaries of the technocrats of the Ministry of Agriculture should be at per with ICAR alongwith promotional avenues, with time scale. Similarly in the States also like the Agriculture Research Scientists scale.

The technocrats felt that administration is negative to them. They give bad P.As., peons etc. and in house-keeping. The senior technocrats are reverted after 5 years as soon as their tenure is over. The replacements are do not planned properly as such the senior posts remain vacant for years. It was pointed out that two posts of additional commissioners were vacant for two or more years, two post of Joint Commissioners in the crops division. It is same with other divisions also. It never happens in the administrative services. They are posted immediately. Now they are being posted in the technical posts on a temporary basis. But why? This raises a very moot point. Are the bureaucrats better than the technocrats and their working? If the officers are not selected ahead of time the work suffers. Should this be allowed to continue? Planning should be done to select the officers for expected post that will be vacant in the coming year.

(viii) **Views of Dr. Mamen**

Dr. T.A. Mamen, Joint Commissioner of Fishers in the Ministry of Agriculture mentioned that "A study group to suggest a comprehensive programme for integrated development of fish farming fisheries activities in the country. In their report submitted to the Ministry of Agriculture, Government of India in February 1989 under the Chairmanship of Dr. A.S. Bose, Advisor, Agriculture, Indian Institute of Technology, Kharagpur suggested the following:

1. *Creation of a Separate Ministry of Fisheries*

Since the very nature of work in "Fisheries" is so different from the rest of the work in the present Ministry of Agriculture or the Ministry of Food Processing Industries, that it required creation of a separate Ministry of Fisheries.

2. *Constitution of an Indian Fishers Services*

Directorate of Fisheries in all maritime states is now headed by Indian Administrative Service Officers. This has a demoralising effect on the technical personnel and serious brain drain has already started. It is therefore necessary to arrest this draw in by creating Indian Services in Fisheries with scales and other benefits comparable to other all India Services.

They said that the State Government were following a policy

which is harmful to the development of Fisheries in the States. Even in some land based states also bureaucrats have been appointed as directors and if this trend continue bureaucrats will replace the technocrats. This is a very undesirable situation. The qualified technocrats in fisheries should be allowed to implement the fisheries programme. If there are shortages in the performance, various training programmes should be organised by the fisheries division of the Ministry and the Directorate of Fisheries of the States to build up the competence of the technocrats.

(ix) Views of Mr. Wakhle

Mr. A.S. Wakhle, Former Member (Services) Telecommunication also mentioned that till 1983 a bureaucrat was the Director-General of Posts and Telegraphs and he used to take policy decisions without taking into confidence the technocrats. In 1985 the Government of India felt that in view of fast technological changes and expansion a technocrat was appointed as the Director-General Telecommunication.

The Telecom Commission was constituted in 1989 and the Chairman, who is a technocrat is the Secretary, Department of Telecommunication (DoT). There is no bureaucrat in the Department except the Secretary to the Commission whose functions are to assist the Commission with Secretariat help.

8

BUREAUCRACY AND BUREAUCRATS

Bureaucrats are a very power group of officers running the administration of the country. Now they are called Indian Administrative Officers (I.A.S.), earlier, i.e., before Independence, they were called Indian Civil Service (I.C.S.). They are supposed to be part of the steel frame running the Government. It must be said to their credit that they did run the administration very well, as the officers were selected with great care, and trained well, both in the Institutions and in the field.

The present system is also of selection and training of I.A.S. officers is good, with a lot of in-service training programmes. They are very good for general administration, as they are trained for that purpose, but they are not good for administration of technical ministries, department, undertaking and institutions.

We have nothing against the bureaucracy and the governments, but we do not think that they should be put in-charge of Scientific and Technical Ministries or Institutions of any kind for various reasons. Their services should be utilised for general administration, ministries and departments of that nature.

There are some inescapable truths about governance in India: the bureaucrat is the real overload, while the politician is at best a transitory phenomenon. The bureaucrat formulates the laws and implements them, while making the politician believe he is doing it all. The bureaucrat's word counts, while the politician's word is forgotten in a few months — if not days. And, if the professional politician did not exist, life would go on.

But in the absence of a professional bureaucrats, India would fall flat on its face. He is the power centre, and knows it, protects himself and his ilk, and does everything he can to perpetuate his supremacy. In India, the *babu* rules. And, the most powerful of the *babus* are undeniably the 6,000 or so, Officers of the IAS.

"This is a very well established Institution," says Bashiruddin

Ahmad, a research professor at New Delhi's Centre for Policy Research. "For many years, the committee of secretaries have been running the Government, simply because many of the politicians don't have the time or the expertise to handle administration."

Any political system which wishes away the bureaucracy is indulging in wishful thinking.

It is a system which has a stranglehold on the entire country from the lower most levels (the fresh IAS) officer who is an assistant collector in *tehsil*) to the very top (the Cabinet Secretary traditionally the godfather of all bureaucrats). Madhu Limaye, a four-time Lok Sabha Member and former general secretary of the Janata Party: "Just 48 hours after the National Front Government took over (in the first week of December 1989), the bureaucrats (IAS) had captured all the top slots in the ministries". The only non-IAS person who achieved a position at the top was Bimal Jalan, Secretary, Finance, courtesy his proximity to V.P. Singh.

Even the junior-most IAS officer has tremendous clout, and some of the manifestations are incredible. "I used to go around the *tehsil* in my jeep," says a senior IAS officer about his days as a greenhorn, "And very often, in a village, people would stop the jeep and make me get out. There would be a *shamiana* of sorts and a microphone, and I would have to give a speech. Imagine giving a speech in Marathi, when you can't speak the language." And of course, he would be garlanded and offered sweets, in much the same way a politician is. "If you go around a district," says P. Umashanker, Director of New Delhi's Indian Institute of Public Administration (IIPA) and an ex-IAS officer (1956 batch), "you will see that the whole world revolves around the Collector. He manages everything – elections, rural development, local conflicts, government policies." As well as law – and order.

Later, when the *babu* has moved up a few rungs as Secretary in a state capital, or as the Deputy Director of a State-level enterprise, he actually formulates policy. Though, as Aiyar says, "Bureaucrat is the 'Instrument of governance' who communicates a policy or a programme to reach whom it is intended for and not someone who actually drafts it". Says Aiyar: "In States like Bihar, where politicians are content to just transfer people from one place to another, the bureaucrats run the show. They decide policy as well as the rules and regulations that direct that policy. That would not do for public consumption." Says Aiyer, "So you create the myth that the minister makes the rules and regulations".

And this myth, and the power it wields, grows with the *babu's* rise through the bureaucracy. The mastery over the magic word "rule" is what a *babu* thrives on, what lies at the core of his power. The Union Minister for energy may say that he wants a power transmission line between Singrauli and Sirsa, perhaps to please his constituents. But the plan has to be processed at the Department of Power, which needs a clearance from the Ministry of Environment and the Ministry of Finance, routed through the bureaucratic levels in those ministries. The *babus* could stymie the proposals for months, if they wanted to.

With independence, the administration ought to have changed. It did not. In fact, it became worse. India adopted a Soviet-style system with its licence-permit syndromes. In essence, favours to be distributed by the New Delhi *durbar* — hence, the additional importance of the IAS officer. A prime example of the licence-permit hold is the steel industry. Forget private products, for them it is a nightmare anyway. Even if a government organisation, say Durgapur Steel Plant, wants to expand its capacity by a million tonnes a year, it has to submit the proposal to Hindustan Steel Ltd. another government agency. From there, it goes to Steel Authority of India Ltd. then to the steel secretary, then to the babus at the Planning Commission, onwards to the Economic Affairs Committee. Ultimately it will reach the PMO, and wind its way downwards. In the middle somewhere, Finance Ministry bureaucrats will step in, claiming mastery — they hold the purse-strings. All this takes months — if not the five-year span of a government. In Japan, the plan would be okayed or negated at a single board meeting.

Another example: Air India says it can finance its aircraft purchases with its own foreign exchange earnings, and handle interest payments on loans in the same way. A quick step, which would ensure that the airline increases its network, with better aircraft, and matches savage competition. But the babus sit fiddling figures and projections till the next century — they have no clue at all about how to run an airline blocking what is an obvious step for decades.

All this may have relevance in a pre-Berlin Wall society — the one undisputed aspect of *Perestroika* is to rid the system of over-government. But how does one do it here in India, where senior bureaucrats publicly state views such as how one of them, and not an army officer, should be the chief of the Army Chief. Or that a babu should be a Commissioner of Police.

Though the bureaucracy is subordinate to the political process — it is supposed to carry out the will of the politician — it has accumulated so much power over the years that it has become an unstoppable behemoth. Says Syed Shahabuddin, an ex-IFS Officer and presently the President of the Insaaf Party: "It's true that we inherited the bureaucracy from the British, but during that period, the politicians were the riders and the bureaucrats the horses. And the rider had complete control over the horse. Today, there has been a reversal of roles. The bureaucracy is the rider and the politician the horse."

But often, as is happening with disturbing frequency, the roles are merged. Says John S. Lall, a former Indian Civil Service (ICS, a precursor of the IAS) Officer: "Fundamentally, in British times, we were responsible to the British Government through a clearly defined hierarchy. We were very close to the seat of authority (now it has a disparaging connotation, he adds). But, on the whole, I think the ICS exercised its very considerable authority with a sense of responsibility".

This, of course, is an ideal bureaucracy, Says Ahmad, the political scientist: "German Sociologist Max Weber used to say that bureaucracy is based on rationality. That a bureaucrat acts in the interest of the state, in a non-political role". This is precisely where the politician steps in, says Ahmad, and "because of his own insecurities, starts to politicise the bureaucracy — by appointing his own men — which can totally distort the non-political function of the bureaucracy." He adds: "You may not have an evil intent, but you appoint them because you want your politician to be handled your way. You want someone who will share your views". And so on, from the Cabinet Secretary downwards, enveloping the entire bureaucrats network.

"Politicisation is inevitable", says Umashanker, the IIPA Director, "It is a reflection of the system. A collector or a district magistrate, for example, is not immune to reality. Simply put, if you do not listen to the politician you can forget about everything: a plum posting, promotions, the power and the glory and, for a distressingly large number of officers, the money. Some people are more ambitious than other," says Umashankar, "at the time of Independence," he continues, "there was a clear line between the politician and the bureaucrat." But this divide is not as clear as it was. An ambitious person may see in the political activity a way to advance his position."

Says Limaye: "They (the bureaucrats) are dominant through sycophancy." And if they are not, they pay.

But the point really is that even if the babu gets flak from the politician, he endures and, ultimately, wins — the system, only grows stronger.

The ICS officers, who faced nationalised flak when the British were their overlords, continued even after the British left. In fact, they probably became more powerful, as only they know the ropes of government and were also the only people who could train a new generation and of Indian bureaucrats.

The big break for bureaucrats came during Lal Bahadur Shastri's stint as Prime Minister. Because of a "lack of confidence", writes Limaye in his book *Cabinet Government in India*, Shastri wanted a strong personal secretary. He introduced L.K. Jha for the job, who took on increasing responsibility and kept expanding the secretariat — thus, expanding the area of control. Shastri's Cabinet Secretary, Dharma Vira, was another powerful figure, who frequently side stepped the foreign service in key issues — he often dictated policy over the Kutch problem with Pakistan, just before the 1965 war.

This leads on to the argument that the cream do not join the services any longer, preferring instead to opt for banking, management, accountancy, advertising. In short: "anything else. There are few creative, innovative bureaucrats these days," says Limaye. The private sector absorbs the professionals. The quality may be affected-though at the top slots no one questions a *babu's* professional competence, after years of operating in the system they are thorough breds-but the organisation is not.

(i) Pay and Perks of Bureaucrats

The bureaucrats are also paid high as compared to the Technocrats. They bring home Rs.2,200 a month when they start their carrier, plus perks big house, a jeep or two, a retinue of servants and the official lordship of a tehsil as an Assistant Commissioner or Collector. By the time and I.A.S. officer becomes a Joint Secretary in Government of India, New Delhi he gets Rs. 6,700 p.m. as salary, a three bed-room house in a good locality and perks besides. A Chief Secretary to a State gets up to Rs. 8,000 p.m. as does a full secretary in a Central Ministry. At the very top the Cabinet Secretary picks up Rs.9,000 p.m. The Fourth Pay Commission has done very well by giving these good scales. But it should be given to the technocrats also of equivalent rank. There should be no difference in pay structure. This difference in pay and allowances create the problem and the disparity become more pronounced with additional perks in their hands.

(ii) **Training of Bureaucrats**

The transition from a virtual non-entity to an all-powerful bureaucrat may seem like an impossibility, but at the Lal Bahadur Shastri National Academy of Administration (LBSNAA) in Mussoorie, it take just 26 weeks of classroom work to turn civil service probationers into power-wielding *babus*. The training is very intensive including prodigal training in districts under Senior Officers.

The rigours of the Union Public Service Commission examinations — a Preliminary Objective test, followed by a main examination and personal interviews — over, a select group of around 300, service probationers make their way to the Academy. They are only from the Group A Services, the Indian Foreign Service (IFS) and the Indian Administrative Service (IAS). The lesser — though increasingly more popular — services, revenue, railways, audit and accounts, among others, train their new officers in establishments scattered across the country (revenue officers, for example, are schooled in Nagpur). The LBSNAA is the place only for the creme de la creme of the crop, which, at the preliminary test stage, numbers more than 90,000.

This training programme, and later on in service training at the Administrative Staff College, Hyderabad and abroad help the bureaucrats to become good officers for general administration, but not for technical ministries, departments, public sector organisations and other technical institutions in general. There are a few exception to this. But this should not be the system and the risk. It should be otherway round. The technocrats and scientists should be given training in administration to become good technocrat-cum-administration.

(iii) **Role of Politicians**

The politicians play an important role in the selection of a technocrat or a bureaucrat of the Secretary of the Ministries in Centre or the Department in the States. So also in the various public sector organisations, corporation etc. Most of the politicians do not have time to go into administration as such they prefer administrators as Secretaries or some time as heads of various organisations. The politicians have to look after their constituency, the district from where they come in the States. Those who are in the Centre they have to look after their big upgrading constituency and the interest of their state as such they tend to prefer some one whom they know and will oblige. But this should not be as all officers be technocrat be bureaucrat will help the Ministers within the rules and parameters of the Government.

Many research workers say that since the politicians feel little insecure in the new changing set up they depend on the bureaucrats and ultimately make them very powerful as has been narrated in the earlier chapter, but this ruins the technical ministries, departments and organisation.

The politicians also have a feeling that if they do not like a bureaucrat, or who does not toe his line then he can be changed easily. Whereas it is difficult to change a technocrat, because the substitute will not be available soon, and where to put the unwanted technocrat. Here technocrats have to pull along with their bosses. This is an art and science.

The powerful lobby of the bureaucrats with a big number wield a great influence on the administration to use that all the top prize posts in all the ministries, departments and organisations are occupied by them. Very often technocrats also help them by their narrow rivalry. This helps both bureaucrats and politicians to play their cards.

9

TECHNOCRATS AND GENERALIST

(i) Views of Administrators and Specialist

The controversy between the technocrats and generalist have been going on for a long time. With the advent of scientific technological development in the world this controversy has come to the surface very widely and pointedly. As our administrative structure was the same as the British, the writing of the British authors reflected on the working of our system also. I have taken views of many British technocrats and administrators who wrote from early sixties till today. The Indian technocrats and administrators have contributed a great deal on this controversy to find out a solution acceptable to all.

James Dunnet wrote that one must admit that whole training of the administrator and the very nature of his job make it right and proper that he should be a cautious animal. To look to the administrator also as the man who should give an organisation its dynamic impulse seems to be putting him on additional responsibility which should not be his to carry: however, his function is to ensure that his department has availability in the rank of experts, men of the quality and imagination to provide this dynamism and it is his function also to create, by advice and sympathy, conditions in which they can do their best work.

Allan S. Murray, a Canadian wrote that the administrator cannot provide all the necessary information for the achievement of the organisation's objectives. He is ultimately responsible for getting this information. On the other hand the scientist can provide essential information for achievement of the organisation's objectives and, therefore, is responsible to the administrator for getting such information. The scientist is not responsible for setting organisation objectives, but he shares with the administrator the responsibility for formulating the plan for his department and is responsible for the functions of his department. The scientist is only occasionally responsible for recommending policy for the whole organisation. He accepts responsibility and authority as a delegated power from the administrator, scientists and administrators must know more about each other's work. Informal training of each by the other could dispel distrust and at the same time promote understanding by each of the other's position, problems and peculiarities. Mutually acceptable and clear lines of jurisdiction and goals or objectives provide motivation for achievement. Salary scales of scientists should be raised so that while practising his own speciality a scientist can reach a salary level equal to an administrator's, if he has proved his ability in his field. Complementary to the foregoing, it is necessary to attach rank to the individual and not to the position. As most system of government do not permit group decision making a programme of well-planned and prepared staff meeting could ensure adequate consideration of the problems. Formal recognition of the areas of responsibility makes the goals and areas of accountability clear and definite. And finally, formal and informal meetings between scientists working in different fields would help develop better working relationship between scientific fields.

R.P. Khosla wrote that the role of the generalist has come to be questioned in recent years for a variety of reasons. Perhaps, the two most important reasons are: first his apparent lack of any professional expertise in a world where professionals are merging in every field of activity; and second, the increasing use of science to achieve social goals. As regards the first one, author defines the areas of professionalization of and the administrator and shows how it is quite distinct from other fields of activity. As regards the second, it can be said that in a world where science is playing an increasing part in administrative decisions what one wants is not scientists who are administrators; but on the other hand, administrators who have a deeper appreciation of the role of science in society. At the heart of the concept of specialization is the fragmentation of issues, the dissension of a problem into its many component parts and the detailed study of each of these parts. When the rapid advance of scientific knowledge is emphasizing this process of fragmentation as an essential precursor to proper study of a problem, it becomes even more important for the organisation to contain within it, people whose particular skills be in their ability to synthesise the news and findings of the experts and decide on the relative importance of conflicting claims. The administration of the technological society of the future, is going to require more than anything else, the professionalised generalist administrator. However, change is admittedly necessary but it has to be a change that increases the professinalisation of the generalist along clearly defined lines. It should not be a change which radically alters an administrative system that already matches the essential needs of democratic government.

B. Sivaraman wrote that administration is like a woman's dress. Administration has to fit the frame of the social and political order of the people to be administered. After examining the merits and demerits of both, especially in the field of science and technology, concludes that whereas the generalist will generally fit in as an overall adviser or coordinator or planner, it is only on rare occasions that the specialist will fit in as such adviser, co-ordinator or planner. It is, therefore, theoretically incorrect to put all services at the same level in system of selection for the top management posts. The generalist having the right training will always be at advantage over the specialist without training because the field is of the generalist.

P.R. Dubashi wrote that administration has become a body of specialised knowledge, skills and techniques with a general application. Generalists are concerned with general administrative tasks in various subject-matter areas. There is a certain amount of transferability of administrative skill and knowledge from one field to another. A few examples are given to explain the need for generalist administrators in specialised fields. If the generalist administrators in specialised fields. If the generalist administrators and the technical specialists co-operate with each other, with malice towards none and respect for all, it would help a great deal in enabling public administration as a whole to reach its goals.

P.C.S. Sarkar wrote, in India after independence scientists and technical experts are playing a vital role in development with no significant change in administrative set up. In relation to the Indian Administrative Service all other Class I services, whether scientific or technical occupy a distinctly inferior position both in respect of pay scales and service prospects. A rational pay structure based on job evaluation is needed to keep up the morale of public service.

Sarkar further examined the relations between the I.A.S. and the specialists (technocrats) in the machinery of Indian administration in the light of the administrative Reforms Commissions recommendations, Pay Commissions Recommendations, and Mrs. Indira Gandhi's Convocation address at the Roorkee University in 1967.

She concluded that the country gets the best from both (specialists and the generalist) and both must strive to equip themselves for future tasks in a world where inter disciplinary approach is becoming the key to advance.

Prabhu Dutt Sharma while discussing the well-known controversy of generalist vs. specialists and concluded that the role of Indian bureaucracy will be largely determined by the levels of professional competence. If the specialists are not inducted in policy-making apparatus of administration in time and various specialisms are not emphasised in sharpening the administrative tools for development, the administrative system in India very soon will avoidably be involved in feuds, detrimental to the interests of the nation at large.

Vishweshwar D. Sharma wrote that the Generalist Administration has failed to keep pace with the fact changing times. The agricultural Reforms Commission too found the administration heavily relying on "unchallenged generalism". The adherence to the concept of the 'gifted laymen' entails most damaging consequences. The prevalent personnel administration condemns the specialists to a perpetual positioning of the side-lines of the administrative hierarchy. The surge of the protests that ensured after publication of the Third Pay Commission's Report, is a clear pointer that the streamlining of the generalist-specialist relationship is urgently called for. The services, a successor of the I.C.S., is more adept in maintaining law and order and managing revenue administration than handling the multifarious task that the development administration involves. The existing bureaucracy must be rationally structured, more functionally oriented and should be "achievement centered". The 'generalist' does not have the type of skills and expertise needed to resolve the steady rising administrative problems, which are often inextricably interwoven with techno-socio-economic complexities. This problem can be solved by matching jobs with the man possessing the needed qualifications in terms of skills, training and experience, etc. The technocrats can be given short-term training in modern management techniques, which will make them better manager than the generalist.

A. Avasthi who made a study of controversy of Generalist vs. Specialist, and the recommendations made by the administrative Reforms Commission, Fulton Commission of United Kingdom, Madhya Pradesh Reforms Commission and referred to various other solutions like the creation of separate hierarchy, parallel hierarchy, joint hierarchy and united hierarchy. He has finally come to the conclusion that instead of distinguishing these two categories as individuals and groups as have been done for. We should in future distinguish between a two tasks and functions. So far as the technical posts are concerned they have to be filled by technicians and specialists. As regards the top policy-making managerial administrative posts, no service or cadre *per se* should have a right to them. There can possibly be no right source of recruitment to them. These top positions should be open to persons who fulfil the requisite qualifications irrespective of their service ground.

N.A. Muttalib said in development administration the passion for increased specialisation on the one hand, and the pressure for greater coordination on the other hand are to be accommodated with a view to replace the present super-subordinate relationship by a new concept of partnership. Indeed, the generalist and the specialist alongwith the elected functionary constitute the "trimwhite" of development administration.

Manohar Lal wrote that the problems of generalists *vis-a-vis* technocrats is deep-rooted. Professionals are unwilling to accept inferior positions. Before the discontentment of the professionals results in paralysing the whole economy, it is desirable that a fair and impartial treatment is meted out to them. Professionals are placed at par with bureaucrats in pay, status, privileges and promotion avenues. Technocrats are not subordinate to generalists. Different sets of essential national services are created, each one having equivalent but parallel grades corresponding to IAS. More technocrats are inducted into technical ministries and departments. Decision making authority is decentralized. Generalists are to develop professionalism, and technocrats who have to administer are imparted on-the-job personnel management training. The education system is reoriented to include personnel management and administration as a regular study by professionals.

Sriram Maheshwari said a measure ascertaining of the specialist is certain in the very logic of modernisation to which India is committed. The demands of the Specialists and the generalist form the point of view of developments and trends in India. A measure of ascendancy of the specialist is certain in the very logic of modernisation to which India is committed. The demands of the specialists are principally two, parity with the IAS in the matter of pay scales and service conditions and, secondly, access to administrative position at the policy-making levels in government. A specialist ought to receive by way of his emoluments as much as, if not more than, a generalist does for the cost of social reproduction of his skills is admittedly high, and his capacity build permanent assets for the society is much greater. But the specialists may not be encouraged to migrate on a very large scale to administrative positions in the secretariat. The specialist positions must be made attractive in terms of emoluments, advancement and status. Also necessary is a planned deglamourisation of the secretariat as a result of which the specialists will find the administrative infrastructure helpful for the practice of their skills in the field and may not be lured to secretariat posts. The same forces that give rise to various specialisms and impart significance to the specialists' role in society call for the stay of the generalist. No specialism today stands complete in itself in terms of the problems confronting the society. Moreover, the specialist judgements must be evaluated in the larger framework of social, economic, political and administrative considerations. These call for consultation, synthesis, coordination, control – the process which are the functions for a generalist. Viewed in this way, both the specialist and the generalist have contrasting, nevertheless complementary roles to play in public administration, "both company are essential for successful administration and the part ought to be seen in relation to the whole."

A.P. Paracex wrote that the role of Generalist in public administration is complementary to each other and not mutually conflicting since both of them have an important function to perform. The officers who join Engineering Services, broadly termed as 'Specialist Services', have to pass through a type of training and experience which give them exposure to a wide variety of managerial tasks. As the engineer reaches the Chief Engineer's level, there is a predominance of coordination and policy formulation role in his functions. In fact he emerges as a Public Works Administrator which is an important field in public administration. This leads to the concept that 'Administration' itself is a specialism and has to be replaced to a particular field of activity. There can be no such thing as absolute administration although we can have economic administration, agricultural administration, health administration, much in the same manner as public works administration. The controversy between 'specialists' and 'generalists' is a fight against a system which denies equal opportunity to a specialist to reach the top administrative levels and thereby denies country's administration the competence available in the specialist services. Fight is against a distorted value system in which a specialist is highly under-rated. The main objective of our system should be to provide good public administration. This can be achieved by implementation of the recommendations of various committees and commissions and the decisions of the Government regarding the relationship of 'specialists' and 'generalists' in the civil service.

K.C. Khanna brought the Bickerings between the generalist and the specialist, central services and the struggle the all India Confederation of Central Government Officers' Association is carrying on an equality with generalist. In development activities the distinction between specialist and generalist has no relevance. The anomaly of according to lower status to specialists in relation to generalist should be removed.

P.N. Agarwal said that as the state now plays a very crucial role in the economic management of the country, in civil service, various new classes notably scientists and engineers have been added but the generalists have remained on top. They are moved too frequently between unrelated jobs. A wider and more important role must be opened up for specialists. They should not be relegated to, play second fiddle. Harmonious and judicious blend of both is necessary for rapid socio-economic development. Classless structure in civil services and administration is urgently needed.

Sakendra Prasad Singh said the generalist-specialist controversy stem largely from the technological and scientific demands of the policy process as also from the differences in pay and status between the members of these two distinct blocs of the civil service. The impact of the controversy has been felt particularly by those countries which have a tradition of generalist dominated civil service such as England and India. That they felt the impact at about the same time is evident from the report of the Fulton Committee of the U.K. (1968) and the Report on Personnel Administration (1969) of the Administrative Reforms Commission of India. Both the reports emphasize that in today's condition the generalist are relevant only to some of the impact functions of society. Some time, engineers, doctors and other specialists, who are engaged in various nation-building activities, are equally important for our society. And hence the demand for parity in pay and status between the generalists and the specialists.

Sisir K. Chaudhary wrote in his article Bureaucrats vs. Technocrats that the specialised work of engineers and scientists is very important in development-oriented administration. Ignoring the prolonged agitation by technocrats will lead to repercussions on development. Bureaucrats are responsible for many of the ills afflicting the public sector. The time-consuming scrutiny of schemes at secretariat level results only in unnecessary delay. With the decline in the quality of political leadership, bureaucrats have been able to frustrate all reform efforts. The disparities in pay and status of administrative and technical service have done harm to our society.

(ii) Views of National Commission on Agriculture

The National Commission on Agriculture wrote in its report that the technical cadres in Agriculture Departments in many States are treated with discrimination, and their grades of pay and status in the bureaucratic structure remain the lowest. Even between technical departments in different States, there are wide differences in scales of pay etc. which come in the way of officials from some States getting posts at the Centre. It is desirable that all States follow uniform pay scales and prospects for the agriculture cadre, which should be comparable to those in the administrative service. All Central technical posts should generally be filled by deputation from the States. The period of tenure should not be too short, and may be fixed at 5 years. In the selection of these officers, their position in the cadre and their competence rather than their pay-scales should be the criteria.

A strong agricultural administration can be developed in the country if an All India Agricultural Service is formed, in which technical officers at the Centre as well as in the States at the level of Class I and above are encadred. This service should have different wings for agriculture, animal husbandry and fisheries. There is already an Indian Forest Service and an Agricultural Research Service. Pay Scales of the administrative and technical services should be at per so that the choice of a candidate of his career is guided more by his inclination and aptitude rather by allurement of pay scales and prestige of different services. There should, in addition, be State Agricultural Services, junior and senior, with provision for inducting competent persons from the senior state services to the all India Service on quota basis.

For ensuring coordination at the highest level, there may be an Agricultural and Rural Development Council which may be set up as an advisory body at the State level under the Chairmanship of the

Chief Minister or Deputy Chief Minister. The Council will consider the basic approach and review general progress of plans.

(iii) **General Views and Actions to be taken**

The controversy of Technocrat vs. Bureaucrat or Specialist vs. Generalist will continue and get worse unless the situation is corrected by giving the technocrats specialists equal emoluments and status with the generalist. This must be given top priority not only by the Government but by also the enlightened public.

Science and technology is developing very fast throughout the world and it is so in India. After Independence India has made rapid strides in science and technology. A large number of institutions have been established by the Government and the public for advance education and research. Every State has got a number of technical colleges and universities giving higher education on technology preparing the students to be very good technologist and administrator. This we must accept.

Similarly in the field of agriculture every State has established a large number of agricultural colleges, animal husbandry and veterinary colleges, agricultural engineering colleges, forestry colleges, horticultural colleges, fishery colleges, home science colleges etc. and top of it agricultural universities for giving specialised education and higher training and research. As such the students coming out of these agricultural universities and allied institutions must get their due share in administration at all levels and more particularly in higher echelons to plan, coordinate and administer the policy of the State and Central Government for proper and speedy socio-economic development of our country. This should be given encouragement, but not denied in one pretext or the other.

All technical posts must be managed by technical persons or scientists as the case may be. As such only technocrats or scientists only should be appointed and not generalist in these positions. The Vice-chancellors of the Agricultural Universities and such allied institutions should be technical persons or scientists only and none else, in the name of administration. Similarly Commissioners of Agriculture, soil husbandry, agricultural engineering, water shed management, soil conservation, veterinary, Horticulture, Fishery etc. should be technically qualified persons with long experience. Similarly the Directors of Agriculture, soil conservation, agricultural engineering, watershed management etc. should be technically qualified with long experience and none else. Unfortunately in some states generalists have been put in charge of these departments. This should be discouraged and stopped.

I would suggest that for proper and speedy technological development of our country in this changing scenario of scientific and technological development, the set up in the Central Government Secretariat and the State Government Secretariat needs change urgently. The technical ministries and departments should be manned by technically qualified persons or scientists according to the ministries or departments at all levels, like department of science and technology and others. At present there are about 14 ministries/ departments in the Government of India which are manned by technocrats or scientists out of 44 are more than 44 ministries and departments. All along the line these technical ministries should be technocrats and scientists except in routine administration under the overall control of the Technocrats or Scientists as the case may be.

In the State Secretariat also all the technical departments should be manned by technocrats only. This is being some States in a few departments. I was additional director of agriculture-cum-deputy secretary (food production) in 1951 when there was no full time secretary of agriculture in Assam and we could implement our food production programme very speedily without any problem. It brought a new dynamism and enthusiasm amongst the officers of the Agriculture Department in Assam at that time. As such when such a system could be implemented in 1951 then why not for speedy socio-economic development of the States and the country as a whole why not how?

In the Public Sector Organisations also only the technical persons should be the head of the organisation.

The systems I have suggested should be tried in all the States and in the Central Government for proper and speedy implementation of the various socio-economic development programmes. It should not be delayed in one pretext of the other.

To achieve the best results of the proposed system which is already functioning in some states and in the Centre, the technocrats and scientists should be trained properly in personnel management and administration. This can be done by introducing one course in each in the final year of the technical colleges in the under-graduate classes and then follow up by courses on personnel management, administration, and financial management after they are recruited to the service in any of the technical departments. The Indian Council of Agricultural Research has already established a well equipped Institute or Rajendra Nagar, Hyderabad, Andhra Pradesh called National Academy of Agricultural Research Management (NAARM). In this Institute little more training should be given on personnel management, financial management and administration, which we lack to a great extent.

The Institute called (MANAGE) National Centre for Management of Agricultural Extension established by the Ministry of Agriculture under Directorate of Extension should also give the above-mentioned training apart from extension education.

The four Extension Education Institute one each at Nilokheri, Anand Rajendra Nagar and Jorhat (Assam) should also organise those above-mentioned courses for middle level officers. The Extension Education Institute at Jorhat has been arranging successfully two courses per year for middle level officers of Agricultural Departments at Jorhat and Veterinary and Animal Husbandry Officers at Khanapara (Guwahati) for all the North Eastern States, West Bengal and Sikki. (9 States). More such courses be introduced.

Every State has an Institute for training of officers, financial by the Ministry of Home. These institutions, should be fully utilised by the States for these courses in a continuing basis.

The Agricultural Universities also should organise such courses for the Senior Officers of the States and also for the staff of the University, including research centres in a continuing basis. This should be given priority.

There are outstanding all India Institutes like the Administrative Staff College at Hyderabad etc. We should take full advantage of these Institutes and send the senior technical officers and scientists for such courses regularly. Similarly there are outstanding institutes in different countries of the world, particularly in United Kingdom and USA. The Generalists are deputed there, but why not the technocrats and the Scientists on a regular basis. This should be worked out by the Technical Ministries in consideration with the Home Minister. Training and continuous reference courses are absolutely essential for improving competence. I strongly feel about it and that is why I have been suggesting to all the states since 1963 onwards when I was Director of Extensive Training. Lots have been done since then. Mr. B. Sivaraman, former Cabinet Secretary and Member, Planning

Commission (Universities) suggested that training of officers in administration etc. is very important for our country.

There should be an officer with supporting staff for working out training policies at the Centre in each Ministry and Department and so also in each department of the State Government. They will work out the training policy and programme and institutes.

They will also provide funds work out when and which officers should go for training one year ahead so that the officers can arrange about their family and other problems if it a long duration course of three months and more. No officer should be allowed to drop the courses. If it is done then this should be entered in the character roll, unless very good reasons are given. We should be very strict.

Each officer who have undergone training should be given an evaluation report by the Head of the Training Institute to the appointing authority for their information.

The training Unit of each university and department in the Centre and States should evaluate the programme and change according to the requirement of the groups.

When all these are done regularly there will be no possibility of any complaint that the technocrats and scientists are not competent to be planner, coordinator and administrative head. They will be the best in the country for their assignment.

Hence all technical posts at all levels should be manned by technocrats and scientists according to the situation. All reports and write ups in book periodicals and papers point out clearly on this direction. As such we must act accordingly without any delay.

10

CIVIL SERVICE AND POLITICIANS

(i) Views of Political Leaders, Administrators and Journalists

The relationship between the politicians and civil service, technocrats, and scientists should be very cordial and understanding. Each other should appreciate the position and the responsibilities of the other. This will help in running the administration smoothly without any misunderstanding or undermining each other's position. The technocrats and scientists will have to learn the art and science of dealing with political bosses. This will require proper training at various levels to be imparted by professionals.

Many political leaders, administrators and journalists have written on this subject as it is very important for us.

D.W.C. Wass wrote that the day-to-day dealings with the minister, civil servants should constantly bear in mind some main factors: first, secure the reputation of the Minister from the slightest hint of in efficiency. Second, the Minister should be spared all but the most necessary requests and in case a submissions is to be made the language used should be concise, simple and clear. Third, the civil servant must learn to accept the need for political expediency as an essential ingredient in the working of a parliamentary democracy. They must remain scrupulously neutral on purely party matters. Their loyalty is to the Government rather than to the party in power. It is essential that facts are presented with the utmost objectivity even though politically unpalatable they may be. The reasoning from the facts should be clear, logical and devoid of innuendoes. Fourth, a civil servant should not hesitate to advise a Minister to reconsider his decision if he feels to be genuinely mistaken. One hopes that Ministers will continue to feel that there is much right about the civil service today and that much of the talk of the need for change is based on a faulty under standing of the role of the civil service. On the other hand, the civil service must not resist desirable change, and it is up to the civil servants collectively to regard the clamour for reform as a challenge to be met. The civil service can justify and prove itself only by approaching the question in such a constructive spirit, and show to Minister, to Parliament and to the public that it is a match for the problems.

Lord Bridges wrote that the essence of the relationship of the permanent head to his Minister is that there can be no clearly defined limits demarcating their spheres of action. The way to make sure that the Minister will have no occasion to say "Why was I not told", lies in an understanding of each other's minds and purposes so close and so intimate that the question can never be asked of the Permanent Head: and it is for the Permanent Head to see that the same understanding is passed on to the senior men in the department. An understanding of this kind between two men is not easily attained. It is not sufficient to disclose decisions taken or schemes in process of formulation. Each must let the other know of purposes and remoter aims for the future not yet clearly defined. Moreover, each must show a sensitivity to the other's point of view so that the exchanges between them bring sympathy and understanding even when they do not end in complete agreement. All this makes demands on qualities not only of intellect, but of character. It is fair to say that the duty of creating and substaining such a relationship rests more heavily on the Permanent Head than on the Minister. The latter is bound to be preoccupied with the wide range of work which his position involves his parliamentary duties, his discussions with the Cabinet colleagues, and his contacts with his constituency and with national bodies. It must be for the Permanent Head, whose working life centres in the department, to see that both he himself and the department as a whole are working in harmony with the Minister's ideas.

J.H. Gilman wrote loyalty generally means the devotion of citizens to a persons, to a symbol of the State or to a system of government. An administrative loyalty, however, takes the narrower from of respect of the policies and decisions taken by the government. A loyal public servant may reasonably take part in the preparation of government decisions and, therefore, considering the public interest, oppose the principles or expected results or even raise technical objections. But once a decision has been taken the official must fully respect the decision and duly carry it out. A mere statement of the ethical principle of administrative loyalty, before and after decisions that may be controversial is not enough. An attempt should be made to measure the civil servants adaptability to political change by means of tests and training courses. Again, the administrative loyalty will flourish only where the political authorities respect the views of civil servants and for their efficiency, civil service career must be based solely on their merits and experiences. Also, the regulations must provide provisions forbidding officials from taking part in any kind of political campaigns and also providing fair and explicit disciplinary measures to deal with any offenders. A better understanding of administrative loyalty and of the public servant's position with regard to politics would wipe off certain difficulties or problems, viz., refusal to enter the public service, unproductive administrative action and rebellion against the established authorities.

R.K. Khadilkar wrote in a parliamentary democracy, the final responsibility for any decision or action rests with the Minister who is accountable to the legislature. From this concept of ministerial responsibility it follows that it is the Minister's privilege to have the last word in policy-making and administration. In practice, it is certainly the function of the Minister to determine policy and it is the duty of the civil servant to make available to his political chief all information and experience at his disposal and to tender advice without fear or favour, so that correct decisions can be made by the Minister. In regard to implementation, it is the duty of civil servant to carry it out without any reservation whether he agrees with that decision or not. In evolving suitable practices to govern the relationship between the Ministers and their principal aides – the secretaries, that as far as possible a minimum record should be kept of discussions on all important matters between the Minister and the Secretary. If there is difference of opinion on an important issue, the Secretary's view as well as the Minister's discussion should be recorded. Similarly, if Minister passes an order which a Secretary considers seriously prejudicial to the public interest from the point of view of the Government of the State as a whole, he should have the right to bring the matter to the notice of the Cabinet Secretary. The public service should not only be responsive to public feelings and sentiments but should also be capable of educating the public opinion and using their discretion to accommodate and even pacify them where need be. The members of the public services should be easily accessible to the public, particularly to the weaker sections, and should be promptly investigate into their complaints and take quick decisions to remedy their grievances. Thus what, in short, is required of administration is, quality, efficiency, initiative and a willingness to help resolve the common man's problems.

B. Mehta wrote, in a democracy, the Parliament representing the people and elected by the people must have upper hand. It is all supreme. But in the interest of smooth running of democracy of which parliament is the custodian, healthy conventions will have to be developed to govern its relationship with the administration or the administration. An equilibrium of mutual understanding is to be maintained where the politician has to learn that parliamentary sovereignty is not legal talisman to bully the administrator and similarly the administrator has to realise that whatever is debated and passed in the parliament is a national mandate and he is nobody to question of sabotage it.

Vishna Prasad wrote that in the days following Independence the relations between the Ministers and civil servants were by and large cordial. But after fourth general elections the problem of harmonious relations between the two added new dimensions. The Congress lost its unchallenged dominance. In a number of States non-Congress Ministers of United Front or S.V.D. were formed. These Governments lacked cohesion and stability. Mad pursuit of power and acute selfishness brought the politicians and Ministers down in the eyes of the people in general and civil servants in particular. The politicians were fast losing the prestige, but the civil servants retained their glory. After Independence their political class did not attract good talents, while the administrative class continued to draw the cream of the youngmen. It is natural that the administrative class may develop an aura of superiority complex. The civil servants know that the Assembly can be dissolved and ministry can be dismissed, but public officials are indispensable. This fills them with a sense of pride. In post-fourth general elections period with unstable cabinets and inevitability of President's rule in the States, such a psychology is natural. A successful administration depends upon constant interaction and a lot of feed-back between the two. One must not try to do the job of other. The celebrated maximum of civil service neutrality must be honoured. The civil servants possess the right to advise, to encourage, to warn and to be consulted. But once a policy is formulated, its faithful implementation is civil servants' responsibility. The political conduct of the Minister is not the concern of a civil servant. Politicians also must realise the worth of the higher administrative personnel. Unwarranted suspicion about the bureaucratic frame should not be given shelter. Political parties should build up a code of conduct to regulate their behaviour. They should build healthy relations with their administrative subordinates.

Shiva Rao wrote that India has copied British practice in several respects in her parliament system, but has ignored the healthy relationship between civil servants and Ministers that has developed in U.K. The problem of relationship between a Minister and a civil servant has come into prominence in several occasions in the past two decades. This has resulted in the steady deterioration in administrative standards. A civil servant may be called upon to do things at the bidding of a Minister which is repugnant to his self-respect previously secretaries dared to express their own opinion, regardless of the Minister' susceptibilities on prejudices. Recently criticisms in Parliament of individual civil servants have led to demoralisation and a sad lowering of standards. Civil servants as a class are gravely concerned about their prospects and security of tenure. The Minister does not always come to their rescue. Democracy can function with a fair measure of efficiency of all political parties see the danger of undermining the morale of the civil services and make a genuine effort to vitalise the administration.

D.N. Banerjee wrote the relationship between the Minister and civil servant should be – and usually is – that of colleagues working together, in a team, cooperative partners seeking to advance the public interest and the efficiency of the Department. The Minister should not be an isolated autocrat, giving orders without hearing or considering argument for alternative courses; nor, on the other hand, should be civil servants be able to treat him as a mere cipher. The partnership should be alive and virile, rival ideas and opinions should be freely considered, and the relationship of all should be one of mutual respect on the understanding, of course, that the Minister's decision is final and must be loyalty and helpfully carried out and that he requires efficient and energetic service.

G.N. Bush wrote generally administrators blame the politicians for most of the ills of body politic and body administrative, and consider them short-sighted, ignorant and not above temptation. This is not correct. The political executives have their political career always on stake and any miscalculation would bring their downfall. The civil servant with his constitutional subordinate position is to advise the political executive and implement the decisions taken even if he is overruled. In many occasions the civil servant may be irritated and even hampered in his work by his political chief. The only way is to try and dispel the clouds of suspicion which engulf the relations between the two. In this regard the civil servant can take the lead.

Charan Singh, former Prime Minister said, though the administrators job is to carry out the government policy, he cannot be asked to do what is illegal or unethical. He cannot take shelter behind a ministerial order, oral or written, to defend an illegal or immoral action. If the politicians and bureaucrats perform their roles honestly and fearlessly there can be no conflict of jurisdiction. Both of them should be committed to public service, honestly and integrity.

Harish Chandra Gupta wrote, the issue of relationship between politicians and administrators in the light of his own experience. The pay scales and pensions of the services are comparatively static, Judge's pay scales also call for upward revision. As laymen take up a lot of time in grappling with technical scheme, technocrats should head the purely technical departments. We are on the verge of economic collapse through political administrative bungling, deficit budgets and unrealistic planning. Unless statesmanship of a high order can set things right soon, complete breakdown may overtake us.

K.K. Katyal wrote, about the relationship between officials and politicians. While the politician finds the bureaucrat unresponsive and wooden, the official feels harassed because of interference. Over the years and particularly during the emergency many distortions have crept into the relationship. It is suggested that a detailed discussion is necessary on the norms that should govern the relationship between the politicians and officials at various levels, to ensure political detachment and continuity in the administration, alongwith integrity and responsiveness of civil servants to popular sentiments.

Om Prakash Motiwal wrote, in a parliamentary system of Government the executive is divided into political executive responsible to the legislature and the permanent executive which maintains the continuity of he administration irrespective of changes in the political set-up. The minister does not possess the requisite techniques of administration, on the other hand civil servants accumulate knowledge and experience to run the government machinery. The civil servants supplement the actions of Ministers without any desire of gaining prominence in the eyes of public, while the political executive may take decisions. The ordinary citizen feels the impact of government through the actions of the government servants who have to administer or execute the orders of government. In India basic feature of he parliamentary system of government is the association together of: (1) an amateur, lay, political, non-permanent, directing body, and (2) an expert, professional, non-political permanent subordinate staff. The former provides the democratic element in administration, the latter the bureaucratic. Political wing makes the government popular and the civil service bring efficiency to it. The successful combination of both the wings runs the administration smoothly. In coalition governments Ministers are drawn from various political parties, unified and clear direction from the government becomes difficult. Policy decisions tend to get delayed and there is lack of coordination in operation measures and the implementation of policies and programmes suffers. Uncertainty of the tenure of Ministry has strengthened the power of higher civil servants.

B.K. Acharya said, that among politicians the search for scapegoats for our present ills is far more widespread then the search for remedies. They attribute most of our troubles to the much maligned bureaucracy. While under British Raj the district officers used to speed more time with villages, their successors are devoting a great deal of their time running around and entertaining VIPs. With the demise of political leaders of stature the administrators are receiving less and less support from their political masters.

Kusum Buch wrote, that as a result of interference in administration by politicians, gradually a feeling has grown that the road to advancement lies through patronage. A stage has come when the system itself has begun to fail, even to disintegrate. Nowhere does the Constitution subordinate the officers of the executive government to the legislature or to its individual members. The legislators by becoming a party to or the instigators of bureaucratic action at the field levels turn the civil servant into an inefficient instrument for implementation. The civil service has to carry out the political will of the party in power but certainly not the whims of individual party members.

K.K. Katyal wrote, that the theory governing the relationship between the politicians and administrators is deceptively simple. In practice while the politician's primary instinct is to try to strength his hold, the administrator's one is to improve his prospects in service. This phenomenon results many a times in tragic consequences. Change in government should not create fear and apprehensions among administrations. This can be averted if the relationship between the politician in authority and the administrators are guided by proper norms. The politicians should not browbeat officials into taking convenient decisions. The officials too should not carry their independence to a stage where it develops a momentum and autonomy of its own breaking loose from the thinking of the policy-makers.

Inder Malhotra wrote, political developments in India since independence have their impact on the relationship between the civil servants and Ministers. If politicians in power have been anxious to bind the civil services to their will, there has been no dearth of high-ranking civil servants to bend over backwards at the former's behest.

(ii) General Views

The eminent persons have pointed out very clearly that the relationship between the civil service and politicians should be cordial, good and should understand each others point or view. We should respect each other and work as partners in administration for the good of the country and the people of India. This will keep the integrity of each other intact. Technocrats and scientists who are part of the civil service in the Government should behave the same way for smooth working.

It is very essential that we, the technocrats and scientists, are given proper training in personnel management, and administration. This particular aspect has been emphasised in the chapter on Technocrats *vis-a-vis* generalist. Regular refresher courses should be organised for different levels of technocrats/scientists-cum-administrators in various institutions dealing with public administration both in India and abroad as it is being arranged now for I.A.S. This must be a continuing process. This will build up the technocrats and scientists to be very good administrators.

Very often political leaders want generalist as secretaries of the Technical Ministries and Departments in the Centre and some other key post and similarly in the States also because they want officers who will carry out their orders as far as possible. If they do not carry out them they can be transferred horizontally to another unimportant post. But by and large most of the generalist manage to stay in their posts for the full tenure as they adjust themselves to the working of their political boss. There is a nexus between the two. This way the technocrats and scientists are kept out from their legitimate promotions very often.

Technocrats and scientists do not generally have posts for horizontal transfers. The political bosses do not get stuck up with difficult technocrats/scientists as secretaries or even heads of divisions or departments. It is also happening in the case of Vice-chancellors of the Agricultural and other Universities. This is a very retrograde and dangerous step. This will ruin the organisations and institutions. This has been expressed very clearly by many eminent persons which I have quoted in this chapter.

At present there are only 14 Ministries/Departments which have Technocrats/Scientists as Secretaries in the Government of India out of more than 44 Ministries/Departments. Key Ministries like Agriculture, Health, Human Resources. Welfare Department etc. are manned by generalists. These should all be under the charge of technocrats, all the line except generalist to deal routine matters.

In the Ministry of Agriculture and Cooperation there are more than 20 Joint Secretaries, 2 Additional Secretaries, 2 Special Secretaries and a Secretary in the Department of Agriculture alone at present. It was not there earlier. Has this improved the working of the Ministry? All these senior generalist officers can be better utilised in other non-Technical Ministries or Departments where they can be better utilised. Most of them are very good officers, devoted and hard-working.

It is time we take a hard decision about the appointment of qualified, competent and experienced officers in the positions like Secretary, Special Secretary, Additional Secretary, Joint Secretaries etc. To achieve this the senior positions of technical officers should be filled up on a priority basis which will serve up to the level of Commissioners in the rank of Additional Secretaries. This is urgent and important.

Similarly in the States also the technical departments should have technical officers as Secretaries, Joint Secretaries, Deputy Secretaries and Under Secretaries. In some States this has been accepted and are functioning as such in some departments. In Assam the Secretary, Joint Secretary, Deputy Secretary and Under Secretaries in the P.W.D. (Civil) and Irrigation Department and Public Health Engineering are all engineers. In Agriculture Department also I was the Ex-Officio Deputy Secretary when I was the Additional Director of Agriculture in nineteen fifties and there was no full time Secretary for Agriculture as it was clubbed with revenue. It worked very well till

I came on deputation to the Ministry of Agriculture in the Centre.

In Sikkim State also the Secretary of Agriculture is a technocrat (agriculture). This will enthuse the Agricultural Technocrats and will bring a very speedy development.

Similarly in Indian Council of Agricultural Research the D.G., I.C.A.R. has been made the Secretary of Department of Agricultural Research and Education (D.A.R.E). But there is dichotomy in administration by not making the Deputy Director Generals and Assistant Director Generals as Additional Secretaries and Joint Secretaries and rest down the line. There is frustration in the vast and important organisation. The system of one line control of Scientists has been recommended by the Review Committee of ICAR and also the Swaminathan Committee of ICAR's Eight Plan. Why there is delay in implementation of these recommendations? This is very frustrating for all the Scientists and technocrats.

The Nalagarh Committee, the National Agricultural Commission, the Bhanu Pratap Singh Committee and others have recommended that Indian Agricultural Service should be constituted as early as possible. Actually it was passed by the Rajya Sabha in 1971, but as the Parliament was dissolved later on the bill could not be placed before the Lok Sabha, thus the bill lapsed. After this no attempt has been made so far. This should be pursued and revived.

Agricultural Research Service has been constituted and pay scales improved. Till the Indian Agricultural Service is constituted the same scales of pay and other facilities may be given to the Agricultural Technocrats in the Ministry of Agriculture in the centre and in the Department of Agriculture and allied departments in the States. This will bring uniformity in the country and will enthuse the technocrats and scientists in the country which will help in speedy development in the field of Integrated Farming System thus increasing productivity in all spheres.

In order to achieve the desired results, to build up the image, excellence, devotion, sincerity the technocrats will have to work very hard and go for continuous training on personnel management, and administration at levels. These are must for the technocrats.

BIBLIOGRAPHY

The Royal Commission on Agriculture in India Abridged Report, 1928.

Report of the Agricultural Administration Committee by Ministry of Food and Agriculture, Department of Agriculture of India, New Delhi, 1958.

Report of the Agricultural Personnel Committee by Planning Commission, India, 1958.

The Report of the Review Committee on Agricultural Universities, Indian Council of Agricultural Research by Dr. M.S. Randhawa, 1979.

A History of the Indian Council of Agricultural Research by Dr. M.S. Randhawa, 1979.

The Directions in Extension Training by Mr. D.C. Misra, November 1990.

Report of the ICAR Review Committee, 1988.

Report of the Working Group on Agricultural Research and Education for the Formulation of Eight Five Year Plan, 1989.

Report of the Committee appointed by the Government of India to advise the State Governments on the Legislation for the establishment of Agricultural Universities in India by Ministry of Food and Agriculture, Department of Agriculture, New Delhi, 1962.

Report on the I.C.A.R. Enquiry Committee by Ministry of Agriculture India, New Delhi, 1973.

Report of the National Commission on Agriculture, 1976.

National Commission on Agriculture, 1978 : Abridge Report by Ministry of Agriculture and Irrigation Department of Agriculture, India, New Delhi, 1977.

Report of the Parliamentary Seminar on Agricultural Administration by Administrative Reforms Commission India, 1968.

Report of the Second Joint India-American Team on Agricultural Education, Research and Extension, Indian Council of Agricultural Research, New Delhi, 1960.

Report of the Seminar on Agricultural Administration by Administrative Reforms Commission India, 1968.

AGARWAL P.N., Role of Generalists and Specialists. *Hindustan Times* (28th May, 1978).

AVASTHI, A. Generalist and Specialist in Indian Administration. *Journal of the Society for Study of State Governments* (July-December 1974).

BIPIN CHANDRA. A Personnel Administration : time for action. *Indian Journal of Public Administration* (April-June 1969).

CHAUDHURY, SISIR K. Bureaucrats vs. Technocrats : Why administrative reform has never taken off. *Capital* (10 November, 1980).

DUBASHI, P.R. Generalist and Specialists : their role in public administration. *Public Administrator* (June 1976).

DUNNETT, JAMES. The Civil Service Administrator and the expert, *Public Administration* (UK) (Autumn 1961).

KHANNA, K.C. Multi-pronged attack on IAS : missing the wood for the trees. *Times of India* (5th April, 1978).

KHOSLA, R.P. Future of generalist. Indian Journal of Public Administration (January-March 1968).

LALL, MANOHAR. Bureaucrats vs. Technocrats. *Public Administration* (India) (February 1974).

MAHESHWARI, SHRIRAM. The specialists and generalist in public administration in India. *Indian Journal of Public Administration* (January-March 1975).

MURRY, ALLAN S. The relationship between the administrator and the scientist in the renewable resource field. *Canadian Public Administration* (September 1964).

MUTTALIB M.A. The anatomy of dichotomy of generalist vs. specialist. *Indian Journal of Public Administration* (October-December 1974).

PARACER, A.P. Role of generalists and specialists in public administration. *Indian Journal of Public Administration* (April-June 1975).

SARKAR R.C.S. Specialists and generalists. *Journal of Constitutional and Parliamentary Studies* (April-June 1973).

SARKAR R.C.S. Specialists vs. IAS Rising Sun (20 July, 1973).

SHARMA, PRABHU DUTT. The generalists and the specialists in administration. In Ramesh K. Arora (editor) *Administrative Change in India*. Jaipur: Aalekh Publishers 1974.

SHARMA, VISHVESHWAR D. Wanted a new philosophy of public personnel administration. *Modern Review* (December 1973).

SINGH, B.P. 'Professionalisation' of the British Civil Service in recent years and its relevance for India. *Indian Journal of Public Administration* (April-June 1978).

SINGH, SAKENDRA PRASAD. Generalist vs. Specialist. ARC and Fulton viewpoints. *Review of Public and International Affairs* (1978).

SIVARAMAN, B. Generalist and Specialist in administration, *Indian Journal of Public Administration* (July-September 1971).

ACHARYA, B.K. Changing the guard new look for the bureaucracy, *Statesman*, 20 October, 1980.

BANERJEE, D.N. Minister and Civil Servant, *Swarajya*, 28 September, 1968.

BHAMBHRI, C.P. *Bureaucracy and Politics in India*, Delhi, Vikas Publications, 1971.

BRIDGES, LORD. The relationship between Ministers and the permanent department head. *Canadian Public Administration*, September, 1964.

BUCH, G.N., Administrator and Political, *Journal of National Academy of Administration*, June, 1972.

BUCH, KUSUM, The decline of Civil Services. *Indian Express*, 22 September, 1980.

CHARAN SINGH, Politician vs. Civil Servant. *Illustrated Weekly of India*, November, 1977.

GILMER, J.H. The Public servant and Politics: Administrative Loyalty, *International Review of Administrative Science*, 1966.

GUPTA, HARISH CHANDRA. Administration : in retrospect, *Public Administration* (India) March 1974.

KATYAL, K.K. Officials and Politicians, *Hindu*, 19th September, 1977.

KATYAL, K.K., Politicians and Administrators, *Hindu*, 4 February, 1980.

KHADILKAR, R.K. Ministers and Civil Servants. *National Herald,* 7 and 8 October, 1968.

MALHOTRA, INDER, Ministers vs. Civil Servants, *Times of India*, 1 June, 1980.

MEHTA, B. Parliament and the administrator, *Journal of Constitutional and Parliamentary Studies*, October-December 1967.

MOTIWAL, OM PRAKASH, Public services in Parliamentary From of Government. *Journal of the Indian Law Institute*, July-September 1974.

PRASAD, VISHNU. Power relations between Ministers and Higher civil servants after fourth general elections. *Indian Journal of Political Science*, July-September 1968.

SHIVA RAO, B., Minister, MPs and Civil Servants, *Opinion*, 24 June, 1969.

WASS, D.W.C. Ministers and Civil Service, *Opinion*, November, 1964.

INDEX